Revised and Expanded

THE CANADIAN
GREEN
CONSUMER GUIDE

Pollution Probe's List of Key Issues for the Green Consumer

In general, the Green Consumer avoids products that are likely to:

- ⚷ endanger the health of the consumer or of others
- ⚷ cause significant damage to the environment during manufacture, use, or disposal
- ⚷ cause unnecessary waste, either because of over-packaging or because of an unduly short useful life
- ⚷ use materials derived from threatened species or from threatened environments
- ⚷ involve the unnecessary use of – or cruelty to – animals, whether for toxicity testing or for other purposes
- ⚷ adversely affect other countries, particularly in the Third World

Revised and Expanded

THE CANADIAN
GREEN
CONSUMER GUIDE

Prepared by
Pollution Probe
(William M. Glenn and Randee Holmes)
In consultation with
Warner Troyer and Glenys Moss

Based on the book by
John Elkington and Julia Hailes

Preface by
Margaret Atwood

M&S

Printed and bound in the United States of America

Canadian Cataloguing in Publication Data

Main entry under title:
The Canadian green consumer guide

Rev. ed.
Includes bibliographical references and index.
ISBN 0-7710-7147-7

1. Consumer Education.   2. Shopping — Environmental aspects — Canada.
I. Pollution Probe Foundation.   II. Troyer, Warner, 1932–   .   III Moss, Glenys.

TX337.C3C35   1991   640'.73'0971   C91–094078–9

Based on *The Green Consumer Guide* by John Elkington and Julia Hailes. First published in Great Britain September 1988 by Victor Gollancz Ltd. Copyright © 1988 by John Elkington and Julia Hailes.

Some recipes in Chapter 3 reprinted from THE NATURAL FORMULA BOOK FOR HOME AND YARD. Copyright © 1982 by Rodale Press. Permission granted by Rodale Press, Inc. Emmaus, PA 18098, U.S.A. Dolphin-Safe logo on Page 35 courtesy Star-Kist Foods Canada Inc. Used by permission.

Cover design: The Watt Group   Cover illustration: Robert Meecham

McClelland & Stewart Inc.
The Canadian Publishers
481 University Avenue
Toronto, Ontario M5G 2E9

CONTENTS

BY NOW, MOST PEOPLE KNOW WE'RE IN DANGER.

We've heard about the thinning ozone layer, the greenhouse effect, acid rain, the destruction of the world's forests, arable lands, and drinkable water. The danger we're in is enormous:

if we don't do something about it, its results could be as devastating as those of a worldwide nuclear catastrophe. We have finally realized that we cannot continue to dump toxic chemicals and garbage into the water, air, and earth of this planet without eventually killing both it and ourselves – because everything we eat, drink, and grow has its ultimate source in the natural world.

However, most people don't know what to do. In the face of such an enormous global problem, they feel helpless. But although the problem is global, the solutions must be local. Unless we begin somewhere, we will never begin at all. An absence of small beginnings will spell the end.

During the depression and the war, conservation was a way of life. It wasn't called that. It was called saving, or salvaging, or rationing. People saved things and reused them because materials were expensive or scarce. They saved string, rubber bands, bacon fat, newspapers, tin cans and glass bottles, old clothes. They made new things out of old things; they darned socks, turned shirt collars. They grew Victory Gardens. "Waste not, want not" was their motto.

Then came the end of the war, a new affluence, and the Disposable Society. We were encouraged to spend and waste; it was supposed to be good for the economy. Throwing things out became a luxury. We indulged.

We can no longer afford our wasteful habits. It's Back to the Basics, time for a return to the Three Rs: *Reduce. Reuse. Recycle. Refuse,* too, to buy polluting products, and *rethink* your behaviour. For instance, use less energy: cut your overhead and increase profits, and stave off a tax hike. Dry your clothes on a rack: humidify your home and lower your hydro bill. Leave excess packaging at the store: let *them* dispose of it. Manufacturers will get the message pretty quick, not just from you but from disgruntled retailers. Start a compost heap. Vote for politicians with the best environmental platforms. Choose non-disposables: razors with real blades instead of the plastic chuck-it-out kind, fountain pens rather than toss-outs. Shop for organic veggies; do it using a shopping basket so you won't have to cart home all those annoying plastic bags that pile up under the sink. Lobby for country-of-origin labels on all food, so you know you aren't eating destroyed Amazonian rainforest with every hamburger bite.

Pollution control, like charity, must begin at home. It's true that industries are major polluters, but industries, in the final analysis, are market- and therefore consumer-driven. If enough of us refuse to buy polluting products, the manufacturers will go out of business. Even a small percentage swing in buying patterns can mean the difference between profit and loss.

This is wartime. Right now we're losing; but it's a war we can still win, with some good luck, a lot of good will, and a great many intelligent choices. This book is a guide to some of those choices. Although they are about familiar, harmless-looking, everyday objects, they are, in the final analysis, life-or-death choices.

And the choice is yours.

Margaret Atwood
Toronto
July 1989

"THE TIME HAS COME FOR HIGHER EXPECTATIONS...

The environment does not exist as a sphere separate from human actions, ambitions, and needs, and attempts to defend it in isolation from human concerns have given the very word 'environment' a connotation of naivety in some political circles... But the 'environment' is where we all live... The downward spiral of poverty and environmental degradation is a waste of opportunities and of resources. In particular, it is a waste of human resources... If we do not succeed in putting our message of urgency through to today's parents and decision makers, we risk undermining our children's fundamental right to a healthy, life-enhancing environment."

Gro Harlem Brundtland,
Chairman of the United Nations World Commission on Environment and Development

Concern about our environment can seem overwhelming. Faced with constant warnings of such global issues as ozone depletion, the greenhouse effect, acid rain, deforestation, and pollution of our air, water, and soil, it would be easy to despair. *What can I do?*

The answer is, quite simply, a lot. All of us have the opportunity — the responsibility — to guarantee a safe environment for our children and *their* children.

It's true a single act by one individual will not end all the problems that plague this earth we share. But as Winston Churchill remarked: "It is better to be frightened now than killed hereafter." Each of us *can* make a difference.

We influence the protection — or destruction — of our environment every day. We do it when we go grocery shopping, when we clean our house, when we take out the garbage, when we take pictures of our kid's first birthday in to be developed, when we drop clothes off at the dry cleaners. If you have the will to make environment-friendly choices, this book will arm you with much of the information you need.

We're largely on our own in this dangerous carnival house of mirrors, jerry-built by our manufacturers, our chemists, our government bureaucrats and politicians. Only by acting as individuals, and groups of individuals, can we ensure our future, and our children's. It means being sceptical and demanding answers. It means reading labels carefully and insisting that those labels are complete and accurate. It means pressing for more environment-friendly farming methods, and consciously working and planning for greater energy efficiency.

There is a relatively simple first step open to us. We think all manufactured chemicals and synthetics should be judged guilty till proven innocent. The notion of "onus" legislation is not new to the traditions of British common law. Anywhere in the world, if asked by a police officer, you must prove your possession of a valid driver's licence by producing it; you are not assumed innocent of driving without a licence until proven innocent. That's an "onus" law — the onus or responsibility to demonstrate innocence is yours.

We have, for a long time, required drug companies to prove their new medications safe before licensing them for sale. The same strictures apply to food additives (though one can and should question the efficiency of regulation, labelling, and most of all, inspection). It's clearly past time to insist on equally rigid standards for the myriad poisonous pollutants in our air, our water, our soil, and our food.

Progress of all kinds demands many forms of evolution. So the processes that saw the demise of the buggy whip and the button hook may well combine to end the production of harmful CFCs, dioxins, and PCBs. The descendants of the buggy whip factory employees, or their neighbours, are probably doing okay today servicing carburetors. The development of the zipper during the First World War did not cause mass unemployment, though it may have discommoded the executive officers and shareholders of the factories churning out trouser-fly buttons.

In blunt fact, no "net" jobs have ever been directly destroyed through environmental clean-ups. Industry has often used the excuse to close geriatric plants that had long since "paid themselves down," as the economists say. But net unemployment always gains in the face of action to save our planet — and our lives. Today — and for as many tomorrows as we preserve — the biggest growth industry is and will be that of environmental protection.

Beginning most dramatically with Rachel Carson's *Silent Spring* in 1962, we've become at least generally informed — and respectably frightened of the grave immediate threats to our ecosystems. First as individuals, now as societies, and ultimately as a global community, we've learned problems exist. So, by definition, we want to help.

With the publication, in late April 1987, of the Report of the World Commission on Environment

FOREWORD

and Development (the Brundtland report, after its chairman), we learned apathy is not an appropriate response to ozone depletion, acid rain, or the greenhouse effect. Our cause is not hopeless: we can help. This decade may offer us our last chance, but it's not too late to save our planet, our environment, our futures. Moreover, said Norway's prime minister, Gro Harlem Brundtland, and her colleagues, it's okay to develop, all right to have aspirations, respectable to grow, if our development is "sustainable." Development (call it lifestyle, if you wish — Gro Brundtland said wisely, "'Development' is what we all do") will work only if it does not diminish, deplete, or destroy elements of our shared environment.

Consider this. *Eco* — as in *ecology*, *ecosphere*, *ecosystem*, and, interestingly, *economics* — is from the Greek, meaning "a household," or, closer to home, "the stewardship/management of a household." As we'll see, the links between economics and ecological health and survival are very clear and real. "Good housekeeping," in short, is what it's all about. We don't walk by an orange peel dropped on the kitchen floor, or scrap paper on the living-room carpet. Nor do we throw trash out the car window, dump toxic solvents down the sink, leave that same car's motor running for 15 minutes while we chat with the butcher or baker, or buy any products in non-returnable or non-recyclable bottles or cans.

The point is that very few if any of our environmental concerns can be treated or understood in isolation. Running our cars, for instance, when we needn't adds substantially to greenhouse gases; it also contributes to acid rain, which washes metals, including aluminum, from the soil. The aluminum, leached from the soil into the surface water, can flow into our lakes, triggering a fish kill. That same metal, scientists now believe, clogs the root system of trees, prevents them from taking up nutrients from the soil, and so starves them until they die.

"The earth is one," Gro Brundtland has observed, "but the world is not." Of course, it *must* be. As surely as an algebraic equation, the world is indivisible. What's done here, wherever we are, has consequences — everywhere. The destruction of a Canadian stream, the poisoning of a water table beneath a pesticide-soaked Quebec farm or a Manitoba landfill site can have both national and global consequences, as surely as a snake bite on one's finger can paralyze the heart.

Left unchecked, our carelessness today, our profligacy, our greed, our pollutants, our waste products will, like Jacob Marley's chains, drag us down and return to haunt our children.

When someone is struck by an auto, or collapses on the street, it may be a bit late for us to learn "the kiss of life," or study how to make a leg splint. Early, "preventive" preparation for environmental first aid is easier, more effective. Let's make a start:

■■■■ Become an informed consumer of the products you buy and services you use.

■■■■ Tell your local shopkeeper to stock the things you want to buy because they're good for the environment. You vote for — or *against* — a healthy world every time you go through a checkout counter.

■■■■ Join an environmental organization and help lobby our governments to do more to preserve the planet for all of our children.

■■■■ Encourage your family, friends, neighbours, and co-workers to do the same.

Warner Troyer
Smoke Lake, Ontario
October 1990

1. COSTING THE EARTH

P

UTTING TOGETHER A SIMPLE PICTURE

of the many different types of damage we are inflicting on the world is not easy. But we must attempt it before looking at what the individual consumer can do to help prevent such damage. Let's approach the task from the outside in.

▲▲▲▲▲▲▲▲▲▲▲▲▲▲▲▲▲

The sun burns 5 million tonnes of hydrogen a second and its core reaches temperatures of some 15 million °C. It radiates a phenomenal amount of energy into space: more than would be produced by 200,000 trillion of our largest existing commercial nuclear reactors.

Earth intercepts only a billionth of the sun's total output, but this is enough to do everything from driving the climate, including the winds and water cycle, and fuelling the growth of the world's crops, right through to burning

▲▲▲▲▲▲▲▲▲▲▲▲▲▲▲▲▲

incautious sunbathers. Indeed, the solar energy entering our atmosphere every year is roughly equivalent to 500,000 billion barrels of oil or 800,000 billion tonnes of coal. This is about a million times more oil than we think there may be left in the planet's proven oil reserves.

Only a small proportion of this energy ever reaches the ground where we jog, walk, and sunbathe, which is just as well. If the sun's raw energy were ever to break through to ground level, life as we know it would be sizzled off the face of the earth. Luckily, much of it is reflected back into space by cloud cover. And the sunshine that does reach the ground is made much less hazardous by something that happens in the upper reaches of the atmosphere.

"We now know that spring is not automatic. We now know that the responsibility is ours to restore and maintain the health of the biosphere."

Pierre Elliott Trudeau (1970)

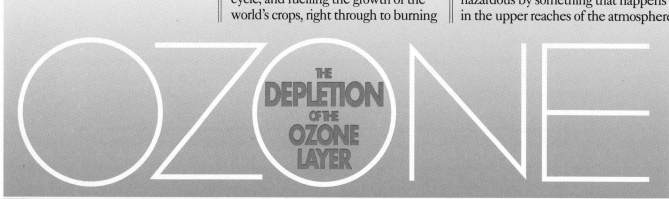

THE DEPLETION OF THE OZONE LAYER

> **❝ It has been estimated that even a 1% reduction in atmospheric ozone could cause 15,000 new cases of skin cancer each year in the United States alone ❞**

Located between 20 and 50 km above the earth's surface, the ozone layer screens out around 99% of the potentially deadly ultraviolet radiation in the incoming sunshine. Yet the ozone layer is so rarefied that if you could compress it to the density of air at sea level, it would be little thicker than the sole of your shoe. Any thinning of this fragile shield inevitably increases the amount of UV radiation reaching the ground.

UV radiation increases the number of skin cancers – it has been estimated that even a 1% reduction in atmospheric ozone could cause 15,000 new cases of skin cancer each year in the United States alone. It increases the number of people suffering from cataracts and other eye diseases and causes extensive damage to crops and other vegetation. It also threatens ocean food chains, because many plankton are highly sensitive to UV radiation. Plankton are the essential food source for many fish and are also important in oxygen production.

The first real evidence that the ozone layer might be threatened was produced by two American scientists in 1974. They warned that synthetic chemicals known as chlorofluorocarbons (CFCs) could thin the ozone layer. CFCs have been used as propellants in aerosols, in fridges and air-conditioning plants (where they serve as coolants), in dry-cleaning solvents, in the plastic foam used to make hamburger and other fast-food cartons, in materials used for furniture stuffing, and in insulation products.

When they were discovered in 1928, CFCs seemed to be perfect chemicals. They were odourless, non-toxic, non-flammable, and chemically inert. Unfortunately, however, they are so stable that they can hang around in the atmosphere for more than 100 years,

slowly drifting up into the stratosphere. Ironically, too, the most useful types of CFC (particularly CFCs 11 and 12) turn out to be the most damaging. Once in the stratosphere, their chemical structure means that they begin to destroy the ozone molecules that protect the earth from UV radiation.

The scientific debate about the extent to which such chemicals destroy ozone raged for years, but eventually a new scientific consensus began to emerge. CFCs, it was concluded by the late 1970s, certainly could damage the ozone layer, but the effects were likely to be less serious than had originally been thought and would be a long time coming. But the chemical industry, which produced nearly 800,000 tonnes of CFCs in 1985 alone, was sitting on a time bomb. That year, British scientists discovered an "ozone hole" opening up over Antarctica.

The evidence had been in American hands for ten years, in the form of data collected by orbiting space satellites, but the computers that process the data ignored the ozone hole because they had been programmed to treat such things as impossible. Once these data were processed into images, it became clear that the computers had been turning a blind eye to an extraordinary phenomenon. The size of the hole varies through the year but can cover an area as large as the United States. Soon Canadian scientists were finding evidence of at least one more ozone hole, this time over the Arctic.

Scientists agreed that the peculiar conditions found at the Poles, particularly the extreme cold and low sunlight for months on end, may have been aggravating the situation. However, the implication was that CFCs could well lead to a global thinning of the ozone layer.

A mass of new research results and growing public concern led to the signing of the Montreal Protocol by

ONE YEAR'S ATMOSPHERIC SOLAR ENERGY

T H E   S A M E   A S

500,000 BILLION BARRELS OF OIL

T H E   S A M E   A S

800,000 BILLION TONNES OF COAL

Canada, the United States, the European Community, and 23 other countries late in 1987. The aim was to cut world CFC consumption 20% by 1993 and another 30% by 1998. In 1990, the Canadian government passed more stringent regulations, which will lead to the complete phase-out of CFCs by the year 1997, three years ahead of the amended international timetable. But, for the foreseeable future, this agreement will simply slow down the rate of ozone depletion because of the longevity of CFCs. Consequently, the destruction of the ozone layer is likely to be in the headlines for many years to come.

* *

THE GREENHOUSE EFFECT

A warmer atmosphere would also cause melting of glaciers and ice-caps, with the result that sea levels could rise by a metre or more.

CFCs are also a minor contributor to another environmental problem whose longer-term impact could be even greater: the greenhouse effect, also known as global warming.

Although between a third and a half of the earth's incoming solar energy is immediately reflected back into space by clouds, the atmosphere as a whole works very much like a greenhouse, trapping heat. When too much heat is retained, we have not just a greenhouse but a hothouse, and the entire global climate system is affected.

The average global surface air temperature has increased by around 1°C over the last 150 years. Within the next few decades, the greenhouse effect could raise the average temperature by another degree, with a rise of several more degrees possible by the second half of the 21st century. Such temperature rises would cause dramatic changes in the earth's climate and weather patterns. The increasingly severe droughts in the River Nile's water catchment area may be just an early symptom.

A warmer atmosphere would also cause melting of glaciers and ice-caps, with the result that sea levels could rise by a metre or more. In the longer term, low-lying cities like Charlottetown, London, Bangkok, New York, and Tokyo could be swamped by the ever-rising tides. Rainfall and monsoon patterns would shift, possibly turning areas like the rice-growing regions of Asia and the North American prairies into dust bowls. Kansas could become the Ethiopia of the 21st century.

These changes in global temperature may not sound enormous, but the earth has not been 1°C warmer than it is today since before the dawn of civilization. In short, we are conducting an unprecedented experiment with our planet.

What causes the greenhouse effect?

C arbon dioxide is the most important "greenhouse gas." It acts rather like a blanket around the planet, holding in much of the solar radiation that would otherwise escape into space.

The level of carbon dioxide in the atmosphere has been growing inexorably since the Industrial Revolution. Between the 1850s and the 1970s, carbon dioxide levels grew by as much as 25%. The main reason for this worrying trend is the ever-growing quantity of fossil fuels (coal, oil, and gas) burned for heat or power.

According to the Worldwatch Institute, humanity added 5.5 billion tonnes of carbon to the atmosphere in 1988 through burning of coal, oil, and gas. If growth in worldwide consumption continues at its recent rate of about 3% per year, fossil fuels could contribute 10 billion tonnes of new carbon annually by the year 2010.

Consumption of fossil fuel is not a transgression committed by industry alone; every consumer must share the blame. Even a relatively conservation-minded household in Canada, without electric heat, uses about 30 kilowatt-hours of electricity per day, about 1.25kWyear/year. If that electricity is all produced by power plants burning fossil fuels, that household would be contributing almost 5 tonnes of carbon dioxide to the atmosphere each year, just through its electricity use.

If that household also uses fossil fuel for heating, it's adding more carbon. A household in southern Ontario can use 1,000 L of heating fuel each season, which would produce 11 or 12 tonnes of carbon dioxide each year. Added to the 5 tonnes produced by electricity generation, that's a total of at least 16 tonnes per household per year; there's no question our domestic energy use accounts for a substantial share of the 50 million tonnes of carbon that Canada adds to the atmosphere each year.

Some politicians and industrialists have suggested that depletion of tropical rainforests is the biggest contributor to global climate change. Perhaps governments find it easier to shift the blame half a continent away. The burning of firewood – and of forests – does aggravate the problem, but the scientific data show that it is our energy use that is the biggest factor. Deforestation contributes much less carbon to the atmosphere than burning of fossil fuel: estimates vary from 0.4 billion tonnes to 2.5 billion tonnes.

S o the notion that we can just plant trees to absorb the excess carbon dioxide is misguided, if well-meaning. Proposals to exchange massive tree-planting for the right to burn more coal and oil are based on a scientific misunderstanding. Trees do have the capacity to absorb carbon dioxide and turn it into healthy oxygen, but that capacity is limited. To grow enough trees to absorb all the carbon from fossil fuels, we would have to grow forests at triple or quadruple their natural density or even more, a feat we have not come close to achieving.

ACID
·R·A·I·N·

Acid rain literally dissolves the surface of stone buildings and monuments and corrodes metal, including cars.

STOP ACID RAIN

Acid rain is distinctly a phenomenon of the industrial age. Throughout North America literally thousands of tonnes of sulphur and nitrogen oxides, the raw ingredients of acid rain, spew forth daily from a variety of sources – the furnaces of coal-fired generating stations, the smokestacks of ore smelters, steel mills, and chemical factories, and the exhaust pipes of buses, cars, and trucks. These industrial activities are all taking place to produce things that will ultimately become consumer items: generating stations produce electricity that we use in our homes and that is used to make other consumer goods; steel mills produce steel for cans, cars, buildings, and bridges; chemical factories produce plastics, pesticides, and more; and the buses, cars, and trucks are used to move goods and people around the continent.

The acids that come from industry and fossil-fuel-powered transportation often rise high into the atmosphere where they can travel with the winds and clouds for thousands of kilometres. Eventually, however, they will come back to earth, either after being washed out of the sky by rain or simply falling out of the sky as dry particles. Even as air pollution, sulphur and nitrogen oxides can have a serious effect on our health. As acids on the ground they have a devastating effect on our environment.

Acid rain was first brought to the attention of most Canadians when Pollution Probe wrote about the damage to our lakes in 1970. Today we know that acid rain kills lakes but also does far more damage.

■■■ The Ontario Ministry of the Environment estimates that as many as 4,000 lakes in that province alone are already unable to support life because of acid rain. Thousands more in other parts of eastern Canada are dead or dying, and European lakes are also suffering the same fate. Within the next 20 years, the $620-million sport and commercial fishery of the pre-Cambrian Shield will suffer a loss of 20% to 50%.

■■■ Trees covering about 4 million hectares of Europe are now showing injury and dying from acid rain. Signs of injury are already appearing in Canada; the maple syrup industry of central Canada, for instance, is seriously threatened. The value of Canada's forests is estimated at $4 billion a year; even a 10% reduction in forest productivity over the next 25 years would be expensive, in resources and in jobs.

■■■ Acid rain literally dissolves the surface of stone buildings and monuments and corrodes metal, including cars. In Greece, six marble maidens that have been holding up a temple on the Acropolis for 25 centuries have had to be removed, their faces almost completely washed away. In Canada, historic marble and limestone tombstones are being erased. The Parliament Buildings in Ottawa are crumbling under the assault of acid rain.

■■■ Respiratory ailments related to sulphur and nitrogen oxides in the air are estimated to be costing Canada $160 million a year.

More than 50 million tonnes of sulphur and nitrogen oxides drift up from sources in Canada and the United States each year. Since 1980 the Canadian government has made reducing acid rain its most high-profile environmental cause, but it was not until 1985 that the government actually started to impose restrictions on Canadian sources of sulphur dioxide. Canada's program calls for a 50% reduction in sulphur dioxide emissions by 1994, and nitrogen oxide levels are being tackled through tougher automobile emission standards. After years of argument and delay, it appears the U.S. is willing to take equivalent measures to control their own acid gas emissions. Amendments to the Clean Air Act passed in 1990 will see sulphur dioxide emissions from coal-fired generating plants and factories slashed almost by half within ten years.

About half of Canada's acid rain comes from the U.S., and many environmentalists are concerned that the American program may turn out to be too little, too late. Even if the American program matches the Canadian acid rain reduction program, no one knows whether a 50% reduction is enough to save our lakes, forests, buildings, and health. Some experts have suggested that we may need to reduce sulphur and nitrogen oxide emissions by at least 80% if we want to reverse the terrible damage that has already been done by acid rain.

Recycling, saving electricity, and reducing automobile use are good ways to help reduce acid rain. It has been estimated that the average household can reduce its contribution to acid rain by 5 tonnes each year simply by recycling its cans, bottles, and papers. Making new products out of old produces less than one-tenth the acid rain created by making the same products from raw materials.

Recycling, saving electricity, and reducing automobile use are good ways to help reduce acid rain.

* *

Deforestation
and Land Loss

The rapid destruction of forests in many parts of the world is one of the contributors to the greenhouse effect and to loss of animal and plant species. Together with bad land use planning and poor agricultural practices, deforestation also contributes to soil erosion. From the tropics to the prairies, we are literally losing ground.

Of an estimated total area of 2 billion hectares of tropical forests worldwide, some 11 to 15 million hectares are lost each year – an area the size of India in 30 years. An area equal to 20 soccer fields is lost every minute.

In 1982, in what was described as the worst ecological disaster of the century, some 3.24 million hectares of forest were destroyed in a fire which swept across Kalimantan, Indonesia. The risk of such catastrophic fires increases as deforestation reduces rainfall in nearby areas.

We tend to think of tropical forests when we think of disappearing forests, forgetting that Canada is also losing forests at an alarming rate. In Canada, we cut an area of forest equal in size to Vancouver Island every four years, just to meet North America's insatiable and environmentally destructive demand for pulp and paper products. Much of the forest cut by the pulp and forest products industries is not properly replanted, some of it is not replanted at all, and some of the areas most recently cut, especially on steep mountain slopes in western Canada, cannot be replanted and will not regrow before the rain washes away most of the topsoil.

In Canada, we cut an area of forest equal in size to Vancouver Island every four years.

As the forests disappear, the pace of soil erosion accelerates. In Guatemala, an average of around 1,200 tonnes of soil are lost every year from each square kilometre of land. As a result, it becomes harder to feed the population, and in countries like India and Bangladesh, the silt shortens the life of dams and can cause widespread flooding in lowland areas.

In 1988 at least 300 people died and over 60,000 were left homeless during "freak" floods in Rio de Janeiro. Brazilian geologists say the floods were no freak. They were caused by the relentless felling of the country's forests — and there is worse to come.

While the global population increases by 84 million people a year, we endanger our ability to feed those people. Each year, for various reasons, we lose 25 billion tonnes of topsoil, enough to cover the wheatlands of Australia. In mid-1989, world grain reserves were at their lowest level since World War II, largely because of the loss of agricultural land.

Although Canada's land mass is huge, only 9% of it is arable, and only half of that amount has climate suitable for agricultural production. Yet we continue to take our farmland out of production: nearly 19% of it was converted to other uses or abandoned between 1961 and 1981.

More than half of Canada's best farmland is within 200 km of Toronto, but we've already built on or paved over 10% of it, and non-agricultural development shows no signs of slowing down. Nor are we taking proper care of the farmland that remains: as much as half of the organic matter in the soil of Ontario, Quebec, and the prairies has been stripped because of monocropping, and we are not putting it back through mulch and manure.

Of concern too is the loss of our wetlands. These marshes, bogs, swamps, and low-lying coastal regions have a crucial environmental role. They protect shorelines, hold back floodwaters, control sedimentation, prevent eutrophication, filter water, and serve as breeding grounds and habitat for fish and waterfowl.

In addition to these natural functions, wetlands are an important economic asset, supporting recreation, fishing, peat and wild rice harvesting, and peatland forestry. Yet many uses are succumbing to urban and agricultural pressures.

In mid-1989, world grain reserves were at their lowest level since World War II, largely because of the loss of agricultural land.

POWER

POLLUTION

Our future contribution to such pollution problems as smog, the greenhouse effect, and acid rain will very much depend on what mix of energy supply technologies we end up using. The options include:

■ **fossil fuels**, such as oil, gas, and coal;

■ **nuclear power**, using today's fission reactors or (possibly) tomorrow's fast breeder or fusion reactors;

■ **renewable energy,** harnessing the sun, winds, waves, tides, geothermal heat, or plants and animals; and

■ **energy efficiency**, which cuts across all of these – and is the key to *The Canadian Green Consumer Guide*'s recommendations on energy.

In addition to our own energy needs here in Canada, we must consider the increasing demands for energy of the developing world. Even if per-capita energy consumption were to remain at its current level worldwide, total consumption would still increase by 40% by 2025; but if people in the developing nations increase their consumption to match Western rates, the increase in total consumption by 2025 would reach 550%, according to the World Commission on Environment and Development.

Even a 40% increase – a scenario that ignores the very real needs and aspirations of the Third World – will strain our collective environment. Clearly we must find effective and equitable means of reining in the ill effects of galloping energy consumption and, ultimately, of reducing its growth.

* *

Fossil Fuels

Oil is the most popular fuel at present, but it is a finite resource. World supplies are likely to be severely depleted within 35 years. Pollution from oil – during manufacture and transport, and in the form of emissions when it is burned – also remains a considerable problem. Supertanker disasters like the spill from the Exxon Valdez may ensure that tanker operations are better managed and policed in the future, but the industry's history of repeated spills around the world does not give us great confidence. Pipelines may appear to be a safer means of transporting crude oil and petroleum products, but pipelines, especially in Canada's north, disrupt wildlife, disturb the fragile vegetation, and are not immune from accident.

Natural gas is often touted as Canada's "clean and plentiful" alternative to oil. But it too is non-renewable and a source of greenhouse gases. Sour gas, natural gas with a high sulphur content, can also worsen local air quality and contribute to acid rain.

A hundred years from now, we will still be using liquid fuels, although they will be very much more expensive. Some of them may well be made from coal, the old fossil fuel standby. There may be enough coal in the world for 200 or 300 years, but its extraction causes considerable environmental damage, and when

Supertanker disasters like the spill from the Exxon Valdez may ensure that tanker operations are better managed and policed in the future, but the industry's history of repeated spills around the world does not give us great confidence.

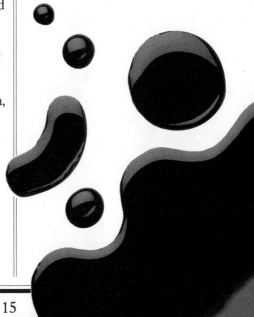

burned, it is often a serious contributor to acid rain (via the production of sulphur dioxide and nitrogen oxides) and to the greenhouse effect (via carbon dioxide).

* *

Nuclear Power

However much damage the burning of fossil fuels may cause, nuclear power remains the least popular energy technology with environmentalists – and with the public. The major problems with nuclear power are reactor safety and waste disposal.

The shadows of the Chernobyl nuclear disaster in the Soviet Union and of the Three Mile Island accident in the United States remind us that even complex machinery with all kinds of failsafe systems can still go wrong through system failure or operator error.

A study by the European Consumer Bureau concluded that the radioactive cloud that spread out from the damaged reactor at Chernobyl carried radiation equivalent to that which would be produced by 2,000 atomic bombs of the size that obliterated Hiroshima in 1945. Some 135,000 people had to be evacuated, and the immediate death toll of over 30 lives is likely to grow considerably as radiation-related diseases surface. The relevant Soviet ministries have been instructed to ensure "the 100% safety of nuclear power plants." Given the experiences of Chernobyl, of Three Mile Island, and of Sellafield in Britain, it would take an ultra-optimist to believe that this is a reasonable target for any country possessing nuclear power.

Nuclear power is often described as the "clean" alternative to fossil fuel. But it too yields undesired by-products – and even ardent supporters of nuclear power tend to draw the line at having those wastes dumped in their own back yards. We have not found a "safe" place for the disposal of nuclear waste.

Large water-filled holding tanks provide temporary storage locations for spent fuel at the power station, but we are a long way from knowing whether permanent safe storage exists for larger quantities of spent fuel and for the decommissioned power plants at the end of their useful lives. Toxic chemicals like PCBs are now a political and environmental nightmare because we allowed them onto the market *before* we knew what damage they could cause and before we knew how to destroy the waste material safely. Nuclear power shows that we still have not learned the lesson.

* *

Renewable Energy

Renewable energy sources are less polluting and won't blow up or melt down, but because they produce energy less intensively than fossil fuel or nuclear power, the facilities needed to capture that energy may be more extensive. A solar energy installation able to produce as much energy as a nuclear reactor might take up to 2,000 hectares, compared with about 60 hectares needed for the reactor plant. It would take perhaps 200 or 300 large windmills to produce the same amount of power as a nuclear reactor, with each group of 25 machines needing an area of around 1,600 hectares. (Also, large windmills, as their neighbours know, can be noisy and may interfere with TV reception.) And if you decide to grow crops to convert into oil, fuel alcohol, or gas, you would need to plant up to 50,000 hectares to achieve the same energy output as a nuclear reactor.

In itself this is no argument against renewable energy, whose role is bound to grow in the future. But it helps to demonstrate that, as you will find throughout

Nuclear power is often described as the "clean" alternative to fossil fuel. But it too yields undesired by-products

this book, many choices that face the Green Consumer involve trade-offs. There is no such thing as a totally "green" means of energy production. But *reducing* consumption – energy efficiency – provides a promising option.

* *

Energy Efficiency

Many Canadian energy experts now regard energy efficiency, or energy conservation, as a "source" of energy. Some estimates suggest that we could easily trim our energy use by as much as 25%, by eliminating energy waste and buying more efficient energy-consuming devices. A 25% reduction in Ontario demand would eliminate the need to complete Darlington, the world's largest nuclear power plant; moreover, we could shut down another two Darlingtons' worth of smaller nuclear plants, coal-fired or oil-fired power stations, or hydro dams. A study by the World Resources Institute suggests that conservation and efficiency measures could cut the industrialized world's carbon emissions in half; presumably, similarly happy reductions in other pollutants would follow, too.

* *

the garbage CRISIS

Garbage is a by-product of a consumer society, and one that we used to regard as unavoidable – until its mountainous quantities began to overwhelm us. Now the problems of what to do with garbage are becoming ever more pressing, and ways to reduce it simply *must* be found. Wastes in the wrong place cause pollution. Finding enough holes in which to dump our garbage is becoming increasingly difficult and less satisfactory as a solution.

Whether we are punching round washers from rectangular sheets of metal, unwrapping a chocolate bar, or removing the packaging from a new washing machine, each of us produces an ever-increasing volume of waste every year. If there were only a few million of us that might be acceptable, but with over 26 million Canadians and around 200 times as many people now living on the planet as a whole, it simply cannot go on.

The galloping increase in consumption has been a largely Western phenomenon until recent years. An average person in the developed world consumes, directly and indirectly, more than 120 kg of paper and more than 450 kg of steel each year; in the developing nations, the figures are 8 kg of paper, 43 kg of steel. Promotion of Western patterns of consumption and waste to the developing world could multiply the environmental damage of the garbage crisis.

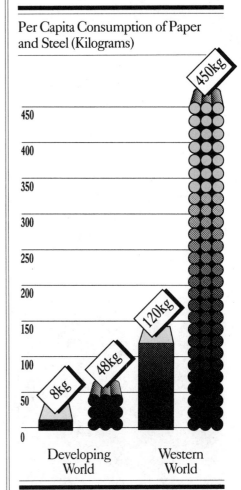

Per Capita Consumption of Paper and Steel (Kilograms)

	450kg
450	
400	
350	
300	
250	
200	
150	120kg
100	
50	8kg 48kg
0	
Developing World	Western World

The galloping increase in consumption has been a largely Western phenomenon until recent years.

Throwaway Lines

■ The average Canadian household throws away one tonne of garbage each year. Canada produces a total of 27 million tonnes of garbage each year, from household, industrial, and other sources.

■ More than 40,000 trees each day are cut down to make the paper for Canada's daily newspapers alone.

■ We discard 1,500 tonnes of steel every day just in food and drink cans. Over a year, that's enough steel to make 350,000 cars – a lineup stretching from Toronto to Winnipeg.

■ Every week Canadians take home 55 million plastic bags from grocery stores. Many of those bags are reused, but eventually they will all be piled into landfill sites.

Garbage is a waste of our natural resources: land, trees, iron ore, and much more. And the manufacture of all the materials we put into the dump has consumed energy and produced pollution. Every tree that is pulped and chlorine-bleached in Canada's mills adds to the dioxin that is polluting the global environment; manufacture of cans and bottles contributes to acid rain and adds greenhouse gases to the atmosphere. The waste that goes into landfill is simply the last remnant of a planet-destroying system that consumes our resources, discharges pollutants, and ultimately threatens human survival.

Chapter 7 of this book focusses on the three Rs of waste management: Reduce, Reuse, and Recycle. Other chapters will show you how to avoid overpackaging and how to dispose of the waste you can't avoid in the most environment-friendly ways possible.

We've spent an estimated \$10 billion trying to clean up the Great Lakes but the menace of toxic chemicals remains.

* *

WATER POLLUTION

We need no better barometer for the state of our rivers, lakes, and oceans than the health of the creatures that live in them.

In the high Arctic, polychlorinated biphenyls (PCBs) and other persistent toxins are detectable in the fat tissue of sea mammals. In the St. Lawrence River, the outlet to the ocean for the Great Lakes, the survival of the beluga whale population is at stake because of massive pollution. Nine of the deadliest chemicals found in the Great Lakes have been identified in tissue samples of dead beluga from the St. Lawrence.

The Great Lakes, our sweetwater inland seas containing 20% of the world's fresh water, are the most compelling example of how we have polluted our waters. Cancerous tumours are found on fish in Lake Ontario. Women of child-bearing age and children under the age of 15 are routinely warned against eating salmon from the lake.

Over 1,000 chemicals used by industry on both sides of the border have been detected in the lakes. Some are highly toxic; others have not been adequately tested to determine their toxicity. Most of the clearly identified hazardous chemicals are not removed by existing water treatment plants, which use 19th-century technology.

Remarkably, there has been little public protest from the 37 million Canadian and American users of Great Lakes water about its quality. We've spent an estimated \$10 billion trying to clean up the lakes, and although we've made some headway – saving Lake Erie from "dying" – the menace of toxic chemicals remains.

So, too, on the eve of the 21st century, do problems caused by inadequate treatment of our sewage. For years, residents of Toronto have been unable to swim at its beaches in the summertime because the level of fecal bacteria is so high.

This disgrace is not limited to the Great Lakes, however. On the east coast, Halifax harbour has been a convenient receptacle for the metropolitan area's raw sewage and now enjoys the notoriety of being one of the most polluted bodies of water in Canada. On the west coast, the Fraser River has been no less abused.

Even Canada's northern waters, so vast that one would think pollution would simply dissolve away in them, are vulnerable to environmental attack. The fragility of our northern coasts was clearly demonstrated by the massive oil spill from the Exxon Valdez in Alaska. In northern Quebec, where the James Bay project has wrought massive environmental change, mercury, a sadly familiar pollutant in Canada, is poisoning native fishing grounds.

Not all water pollution is caused by the discharge of substances directly into water bodies. The most salient example of water damage from airborne sources is, of course, acid rain; across the Canadian Shield, lakes are succumbing to it and no longer bear life. It is now recognized that industrial and urban air pollution swept out to sea is also a major contributor to some forms of ocean pollution.

For too long, our approach has been one of "out of sight, out of mind." But change is now very much in the wind.

* *

ENDANGERED SPECIES

Extinction is a natural process. Well over 90% of species that have ever lived on Earth have disappeared. Many were replaced by others that were better adapted to changing environments, although some disappeared as the result of massive natural disasters.

The appearance of the human species began a significant acceleration in the average extinction rate, however, as we hunted for food, commerce, and sport and converted entire landscapes into increasingly controlled farm and city scapes.

By the early years of the 20th century, roughly one species a year was being lost to extinction, but the pace of environmental degradation and species destruction has since taken off at an alarming rate. In the 1980s, we are losing perhaps one species a day from the 5 to 10 million species thought to exist. We may lose another million species by the end of the century. In 50 years more than half of all species will be gone if present rates of extinction continue.

Without our realizing it, our consumer choices can sometimes tighten the screw on endangered animals and plants.

Each species we push into extinction is like a thread pulled from the tapestry of life. You can pull a fair number of threads out of a tapestry without appearing to affect it, but then whole sections fall to pieces.

As consumers, we may be encouraged to buy a range of products made from animal or plant products derived from endangered species. Without our realizing it, our consumer choices can sometimes tighten the screw on endangered animals and plants.

While most of us may now be aware of the threat to species like the tiger, gorilla, elephant, or rhinoceros, the threat to many other species is not yet widely recognized. Following consumer boycotts during the 1970s,

whale products are no longer used in cosmetics, although oil from the endangered basking shark is. Endangered reptiles may turn up in the form of handbags, wallets, purses, belts, or suitcases. Sea turtles are turned into stuffed specimens, shells, soup, oil, combs, or jewellery.

Many of us know of the threat to cacti and orchids, but few of us realize that even plants like the cyclamen, widely sold in garden centres, are now endangered in the wild. As long as it is cheaper to uproot wild plants than to artificially propagate the species, some people will continue to raid the threatened wild resources of countries like Portugal, Spain, or Turkey. Guidance on some products to avoid is given in later chapters.

* *

Animal Welfare

The way we treat animals is a reflection of the way we treat each other, and it seems the cycle of violence is continuous.

Animal protection is an issue that the Green Consumer can hardly overlook, emotionally charged though it may be. The debate about the acceptable limits to the use or abuse of animals encompasses moral, legal, and economic issues too complex for this book to set out in detail. But we can point out instances where the consumer can make choices that have implications for animal welfare.

According to government figures, 440 million animals are raised in Canada

each year for consumption. That's 16.5 animals per Canadian! Each year, too, over 2 million animals die in Canadian laboratories, says the Canadian Federation of Humane Societies (CFHS). Founded in 1957, the CFHS acknowledges that the Canadian Council on Animal Care has brought about "major improvements for lab animals" but "we need to go much further." In 1989 the federation formed a legislative working group to develop and press for national legislation for the protection of lab animals. They are also devising "codes" of practice for handling food animals.

There are signs of changing public attitudes towards animals. In late

1988, three young grey whales trapped in the Arctic ice were rescued by international effort, to the approbation of a world watching it on television. Only a few decades ago, the grey whales were being slaughtered by North Americans and Europeans; public pressure brought that hunt to a close.

Leisure activities whose aim is to observe animals rather than capture or kill them are growing in popularity. Birdwatching is one of the fastest-growing pastimes of North Americans. Increasing numbers of travellers are going on safari to the Amazon and Africa's Rift Valley. They carry cameras and identification books rather than guns.

Issues of animal welfare have less to do with promoting animal rights and more to do with restricting human rights. Pain and suffering are cruel facts of life. We can never realistically hope to truly eliminate these from our own lives or from the lives of animals. But the pain and suffering we deliberately inflict is a different matter entirely — because we can control it and prevent it from ever taking place. Issues of concern for animal protectionists include factory farming, cosmetic and product testing, pound seizure, fur, and zoos.

* *

WITH SUCH A MASS *of serious environmental dangers surrounding us, it is tempting to throw up our hands in despair. But despair can only make things worse. We can save the planet if we all take responsibility for it in every aspect of our daily lives, starting now.*

GREEN CONSUMER POWER CAN MAKE A DIFFERENCE.

2. FOOD and DRINK

FAR MORE THAN BREAD IS THE STAFF OF LIFE.

Food and drink in general are as close to home as we can get. Yet as with much else in our world, things have gone very wrong in the supply of these vital ingredients that we and our children consume every day of our lives. Indeed, Canadian studies have shown that between 85% and 95% of all our exposure to poisonous chemicals in the environment comes through our food.

▲▲▲▲▲▲▲▲▲▲▲▲▲▲▲▲

"Our fathers and ourselves
sowed dragon's teeth
Our children know and suffer."
Stephen Vincent Benet

▲▲▲▲▲▲▲▲▲▲▲▲▲▲▲▲▲

Personal health considerations aside, the synthetic chemicals used to produce our food have enormous potential to damage our entire environment, including the air we breathe and the water we drink. Much damage has already been done. And although ever more food companies are realizing just how valuable a good environmental image can be, simply removing all additives from a product, proclaiming by inference that it is now "good" for us, is not the answer.

To be truly environment-friendly, the company must find a source of food that isn't contaminated by pesticides and similar chemical residues – which are *not* additives in the legal sense and do not have to be listed on labels.

> **"The grain needed to provide a family of four with just one serving of hamburgers could feed someone eating a grain-based diet for more than a week"**

THE WAY WE FARM NOW

Most of the food we eat today is produced with the aid of vast amounts of synthetically produced chemicals. Where once farms were "mixed" and all-round farmers worked with nature to produce a variety of crops and animals, there is now "monoculture." Farms produce only one or two types of crop, year after year, and most managers of this industrial farming are single-minded specialists to whom the use of chemicals to extract everything possible from the land is now the natural order of things. However, less than 0.1% of most of those chemicals actually reaches their intended targets. The remaining 99.9% contaminates our air, water, and soil and, of course, other food.

Those who work to grow the food we eat are in most immediate danger. The world's use of pesticides has gone from almost nothing in the mid-1940s to about 2 million tonnes a year. It's estimated that three people are now poisoned by pesticides every minute. Improper handling and application of fertilizers result in an estimated 10,000 deaths per year in the Third World.

Here in Canada, a 1983 Alberta Department of Agriculture study reported that 10% of the province's grain farmers showed symptoms of pesticide poisoning — and that 90% were concerned about such poisonings in general.

* *

Conventional Farming: What Is It?

In one recent year, Canadian farmers used 88,000 tonnes of pesticides to kill insects, weeds, and plant diseases, on which they spent a total of $869 million. That amount bought 5,000 different chemicals. Many of them were systemic toxins that penetrate the skins of fruits and such vegetables as lettuce and can't be washed off. In addition, some pesticides degrade slowly and can appear in foods years after the chemicals were last used in the fields. Fortunately, in Canada, many such pesticides have been banned, and those currently in use are much less persistent.

On a more encouraging note, a study released by Agriculture Canada's Agri-Food Safety Division in February 1990 revealed that Canada's food supply is freer of many drug residues and pesticides than even some of the experts expected. The report shows that most meat, fruit, and vegetables sold in Canada complies with federal regulations governing acceptable limits of antibiotics, hormonal preparations, and other drugs and chemicals.

We have sophisticated scientific equipment than can measure a food's pesticide residues to the most minute amount, but no equipment that can tell us what the long-term effects of eating that food will be. Current legal limits for pesticide tolerances, which many consider to be too high, are based on food consumption that's considered average for adults. Not only are the averages sometimes sus-

Most of the food we eat today is produced with the aid of vast amounts of synthetically produced chemicals

pect, they make no allowances for adults who eat more of a food than the average, and they make no allowances for children. Where fruit might be considered 20% of an adult's diet, it can be 34% of a preschooler's.

Nor can any scientific test yet tell us the results of eating a combination of pesticide residues. A Nova Scotia study on Canadian produce found that 60% of the strawberries and 50% of the celery tested showed traces of up to 39 pesticides.

****

Circle of Poison

Both Canada and the U.S. have regulations about what pesticides are allowed for use in their countries. Although Canada does not, the U.S. exports pesticides recently banned in their country to less developed countries for use there. Typically, workers in these countries do not have proper protective equipment or any training in applying pesticides. When produce is imported from these countries, we buy back those banned pesticides in our food. And so the circle of poison is complete.

There are problems far beyond pesticides, as well. Rather than being treated and recycled into the soil, farm manure may be discarded or improperly applied as fertilizer, seeping into waterways where its nitrogen content poisons fish and, possibly, people. The soil in the fields erodes and disappears with the wind and the rain because of its low humus content and today's planting practices. A centimetre of topsoil takes about 150 years to form naturally; we're losing it at the rate of about a centimetre every four to eight years.

We can no longer assume that the world's supply of food will continue to increase. Not even our grain crops and stockpiles, a prime indicator of global food security, are growing. And according to one Saskatchewan farmer referring to the use of synthetic chemicals (inputs), "In conventional agriculture, it's inputs that kill you. You have to keep putting higher and higher inputs in." In the late 1960s, every additional tonne of fertilizer used in the U.S. corn belt could add 15 to 20 tonnes of grain to the world's harvest; now it may add only 5 to 10 tonnes.

Despite this, about 450 million tonnes of a world harvest of 1.5 billion tonnes of grain are fed only to livestock, primarily beef cattle. There are few more inefficient ways of producing food: the grain needed to provide a family of four with just one serving of hamburgers wolfed down in minutes could feed someone eating a grain-based diet for more than a week.

Today's type of farming gobbles up huge amounts of energy, too. For each square kilometre of land, U.S. farmers use about 5 tonnes of fuel a year; European farmers about 12 tonnes. In food production, Canada is among the most energy-intensive countries in the world: nine units of petrochemical energy go to produce just one unit of food energy.

In addition, the burning of these fossil fuels and of wood releases large amounts of nitrous oxide into the atmosphere, contributing to the greenhouse effect. And so, for that matter, does the chemical nitrogen in artificial fertilizers. Less than 50% of these fertilizers are taken up by a crop; some of the nitrogen is also converted to nitrous oxide in the soil and escapes into the atmosphere, where it contributes to acid rain.

Those fertilizers, along with pesticides, contaminate more than our air. From the soil they seep into groundwater, the source of water that lies underground. In rural areas, groundwater may be the only source of well water fed to people, livestock, and crops. Contaminants from groundwater also seep into lakes and rivers and occasionally even bubble back to the surface of the ground. In a sense, we are recycling our own poisons.

We're growing food from oil not soil

THE ECOLOGICAL ALTERNATIVE

The 1991 Canadian Directory of Organic Agriculture, published by **Canadian Organic Growers**, provides a country-wide listing of organic producers, distributors, retailers, and suppliers of everything from organic farm and garden fertilizers to bed-and-breakfast accommodations serving organic fare. The directory is available from COG, Box 6408, Station J, Ottawa, ON K2A 3Y6.

Our great hope for a food supply that is safer for both ourselves and our planet is the alternative type of farming called "sustainable agriculture." It includes the farming systems referred to as organic or biological agriculture, bio-dynamic agriculture, the French intensive method, and ecological agriculture. All share the common goals of stressing the quality of food and of growing it in sympathy with natural processes. We'll use organic farming as an example.

There are six basic standards drawn up by the International Federation of Organic Agriculture Movements. Briefly, they call for the following:

▬ that an organic farm draw upon local resources instead of using outside material;

▬ that the organic farmer maintain and improve the fertility of the soil instead of depleting it;

▬ that the organic farmer avoid any form of pollution when raising and harvesting crops;

▬ that a high nutritional quality in food be emphasized, as well as quantity of food;

▬ that the use of fossil fuels like oil be kept to a minimum; and

▬ that the organic farmer offer satisfying and financially rewarding employment for farmworkers.

You will find, then, no synthetic fertilizers or pesticides, no growth hormones or livestock feed additives, on an organic farm. Instead there is self-sustaining soil that has a high humus content and suffers minimal erosion, producing plants less susceptible to attacks from diseases and pests. Plant residues and compost also help to build a more fertile soil full of earthworms and bacteria; the aim of an organic farmer is soil that gets a little richer every year.

Crop variety and rotation are important on an organic farm because growing just one crop year after year is an open-house invitation to insects and to the quick spread of fungi. This type of farming is a closed, regenerative system where nature's cycles and recycling take precedence. The extractive type of system, on the other hand, uses up soil and doesn't replace what was taken away.

Animals benefit from organic farming as well. They're not penned in, and they're not routinely given antibiotics or other drugs to stop the spread of disease. There is usually so little disease, indeed, that vet bills may be non-existent. Livestock help to recycle nutrients too: fed the organically grown vegetables and grains, they in turn provide manure that will eventually enrich the soil.

Not surprisingly, organic farms tend to use more labour than other types. It's time-consuming just to observe changing conditions in the fields, to feel the soil and see "how many earthworms are in a handful," to compost and to hand-weed. "We suffer badly from the weeds," says a Manitoba farmer quoted in a Canadian Organic Growers study. "But so do the people who spray."

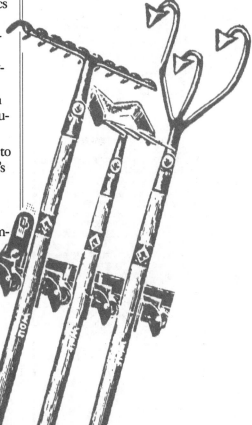

Shopping for a Food Store

One of the best things you can do for yourself and your family is to talk to the manager of your grocery store or supermarket about the types of food you'd like to be able to buy and the practices you'd like to see followed. Take a few minutes too to write a postcard or letter to the president or head office of each supermarket chain in your area (ask the local manager for the address). Often it takes only a few comments from concerned consumers to make retailers start thinking about changing their strategies, especially if they must do so to remain competitive. And see how many "yes" answers your store rates on the following aspects:

Bags
- Does it offer a choice of different kinds?
- Does it charge for bags?
- Does it provide rebates for bringing your own bags?
- Can bags be returned for recycling into new bags or other products?

Packaging
- Does it have a meat counter where you can get products not wrapped in plastic?
- Does it avoid plastic-wrapped vegetables?
- Does it have a bulk section and, if so, can you bring your own containers?

Waste
- Does it recycle its cardboard containers?
- Does it have a recycling program in its offices?
- Does it put unsold food to good use, by giving it to charities or by sending it to be composted?

Products
- Does it offer certified organic food?
- Does it offer alternative cleaning products (see Chapter 3) in the cleaning products section?
- Does it label country of origin on its produce?
- Does it boycott California grapes?

Cleaning
- Does it refuse to use chemical extermination techniques?
- Does it use non-toxic cleaning products on its floors and fixtures?

Promotion
- Does it advertise any of the above with the reason why it's important?
- Does it have in-store promotional material and announcements about environmentally beneficial consumer products?

In recent years some Canadian supermarket chains have introduced, or announced plans to introduce, products they deem "environment friendly," "natural," or "green." Consumers should be aware that such products can be more a matter of marketing and packaging than of substantial benefit to the environment. Make sure that purchases are truly a matter of substance over style.

The Demand for Organic Food

More and more Canadian farmers are switching to organic methods. (Some, of course, have always used them.) The Quebec Ministry of Agriculture, for example, estimates that the province has 2,000 farmers who've converted to organic practices or are planning to do so.

Since we buy a great deal of our food from the U.S., it's interesting to note that American organic farmers already share an estimated $5-billion portion of that country's $36-billion annual fruit and vegetable market. Consumer demand is growing rapidly in North America; in Canada it is so heavy that many organic producers sell everything they can grow at the farm gate or their front door. Quebec's estimated demand is ten times the available supply of most products.

And consumers are also prepared to pay more for organic food. A mid-1988 consumer survey found that even 70% of those who don't buy organic food at present would be interested in doing so if the products were easily available. Not only that, 53% of all those polled said they were willing to pay up to 25% more for organic foods.

Most of the extra cost of organic products is due not to production costs but to the lack of economies of scale in processing, transportation, and retailing; larger farm operations can get their products to buyers for less money per item. And even if our environmental problems were solved, organic farming is never likely to be done on the scale achieved by large corporate farms.

That being the case, many farmers are considering using at least some of the techniques of organic farming, including Integrated Pest Management (IPM). It relies on ecological control of pests, resorting to chemical treatments only if that defence fails. Among IPM's goals are the use of tactics that are most in harmony with human and environmental health and most conserving of non-renewable energy fuels.

* *

Organic Certification

Each province has its own independent certification association for organic produce and meat. No national body exists. Lacking government control, this means that the food industry is "self-regulating." However, because of the increasing use of terms like "organic" and "natural," Consumer and Corporate Affairs Canada and Agriculture Canada are considering drafting more binding legislation on their use, as well as establishing a national certification body.

According to the Organic Food Producers Association of North America (OFPANA), "certified organic" applies to produce grown from land that has been free of all synthetic chemicals for at least three years. Only natural alternatives may be used to enhance growth, control plant diseases, and inhibit predators. California farmers must meet far less rigid standards; they can qualify for a "certified organic" label when their fields have been free of chemicals for only a year.

At Consumer and Corporate Affairs, "organic" refers to how food ingredients are raised, processed, and preserved. To be labelled organic, crops or animals must be raised without synthetic chemicals and a finished food product must have been made without non-organic ingredients and additives.

Any Canadian farmer wanting certification must undergo stringent inspection and accept rigid production standards. There are at least 39 different organic certifications worldwide, but OCIA International (it also operates in the U.S., Mexico, Argentina, Peru, Belize, and Turkey) has been successful in developing generally agreed standards in North and South America, and hopes to have them accepted around the world.

* *

Living with Blemishes

Anti-pesticide consumers say "Beware of perfect-looking produce," because it is likely to contain high chemical residues. And indeed organic food may be less symmetrical, a bit uneven in colour, with the occasional bruise, spot, or other blemish. But any cosmetic defects are just that – cosmetic. Don't confuse good looks with character.

* *

When buying organic food...

look for a certification label from one of the following agencies. Remember that their standards may differ when it comes to what may be "certified organic," so you may want to write for a copy of their guidelines.

California Certified Organic Farmers (CCOF)
P.O. Box 8136
Santa Cruz, CA 95061

Canadian Organic Crop Improvement Association (OCIA)
35 Alexandra Blvd.
Toronto, ON M4R 1L8

Consumer and Corporate Affairs Canada
601–4900 Yonge St.
Willowdale, ON M2N 6B8

Similkameen Okanagan Organic Producers Association (SOOPA)
RR1
Cawston, BC V0X 1C0

Society for Biodynamic Farming and Gardening in Ontario ("Demeter" label)
RR 3
Acton, ON L7J 2L9

If food is being offered as "organic" but you don't see the label, ask your grocer. If you don't get satisfactory answers, contact:

Organic Foods Production Association of North America
c/o Ecological Agriculture Project
Macdonald College
Ste-Anne-de-Bellevue, PQ H9X 1C0

* * * * * * * * * * * * * * * * *

- Support Canadian farmers by buying Canadian produce.
- Sustain organic farming by buying organic produce.
- Press the federal and provincial governments to provide support to farmers interested in adopting sustainable farming methods.
- Lobby the federal government to adopt a national soil conservation policy and a federal system of certifying organic produce.

* * * * * * * * * * * * * * * * *

With an ever-increasing demand on food banks, it is clear than many people cannot afford current food prices, let alone higher-priced organic foods. It is important, therefore, for those who can afford it to support organic farmers whenever they can. As the demand for produce increases, prices will decrease, and organic produce will become more affordable for all — consumer and farmer.

There is a continuing and passionate debate over both the safety and the wisdom of irradiating food.

Sick and Vanishing Land

There is no question that we are threatening our food supply by debasing our environment. Soil degradation is an enormous problem everywhere in the world: an estimated 25 billion tonnes of topsoil is lost every year through erosion.

In Canada B.C.'s Fraser River valley was losing up to 30 tonnes a year per hectare in the mid-1980s; Nova Scotia's cultivated land, 2 to 26 tonnes. Prince Edward Island has already lost half its topsoil since the turn of the century.

Soil degradation includes erosion by wind and water, soil salinization, and oxidation. It costs Canadian farmers more than $1 billion annually, with up to $400 million of the damage due to water erosion, $300 million due to wind. The economic impact from prairie erosion is $380 million a year and growing 5% a year. To complete the cycle, soil degradation contributes to the deterioration of surface water and groundwater, passing along residues of chemicals that may have increased crop yields but did not prevent the land from eroding.

An equally alarming problem is the rate at which Canadian farmland is disappearing under concrete. Only about 10% of Canada's land can support food production; only 0.5% is rated Class 1 farmland (suitable for most crops). More than 50% of the latter is in southern Ontario, and the Region of Peel alone lost 15,700 hectares of primarily Class 1 farmland between 1971 and 1986, about 3 hectares a day. Yet Ontario is a net importer of food, relying increasingly on foreign foods that could be grown in the province but aren't.

In Canada as a whole, a quarter of a million hectares of rural land was converted to urban use between 1966 and 1981, 57% of it "prime" (Classes 1 to 3) agricultural land. And about two-thirds of Canada's considerable food imports are now of products that are grown here already.

* *

THE IRRADIATION OF FOOD

The next thing we're going to have to come to terms with in our food is irradiation. Briefly, this is a process of bombarding food with radiation to preserve it. Given crop losses of up to 50% from spoilage in tropical countries, and the lack of refrigeration facilities, it sounds like a good idea. It has some drawbacks, however, as you might expect.

First, though, it must be said that irradiation does *not* leave food radioactive; the food never touches radioactive material. The process developed in Canada uses gamma rays from cobalt-60 in much the same way as they are used in radiation therapy for cancer. As in that therapy, the dose of radiation is designed to suspend the growth of living cells or destroy them. It can preserve or disinfect foods, as well as prevent sprouting and kill pests like flour weevils.

Irradiation has been used in our health care industry for more than 20 years; by the mid-1980s, almost half of all disposable medical supplies were being sterilized by the process. Irradiation of potatoes, onions, wheat, and flour has been permitted here for about 25 years, although it has never been done commercially, and no irradiated food has ever been sold here. In May 1987, a parliamentary committee said there were still too many troubling and unanswered questions about the potential hazards. Yet in early 1989 the federal government decided to make irradiation more commercially attractive to the food industry. Health and Welfare Canada declared irradiation a food process under the Food and Drugs Act, rather than a food additive, meaning it is subject to less stringent control. Critics suggest that the main reason for the decision was to keep our faltering nuclear industry in business – few countries are buying nuclear power stations these days.

The international unit of measure for an absorbed dose of radiation is the kilogray (kGy), and 1 kGy equals 100 kilorads. The range of irradiation normally used with food now ranges from 15 kilorads for potatoes to as much as 10 kGy to completely sterilize meat and poultry. Up to 1 kGy is a "low dose" that inhibits sprouting, delays ripening, and helps destroy insects; 5 kGy, a "medium dose," kills some bacteria and so extends shelf life; 10 kGy, a "high dose," will destroy viruses and totally sterilize meat and poultry so that they will keep with no refrigeration.

Health and Welfare sets no maximum permissible dosage for human consumption. Under Canada's new regulations, "efficacy tests" will be required to determine the level of irradiation for which an applicant wants approval. No new foods have been added to the approved list. The most likely early candidates will be fresh poultry, fish and other seafood, fruit, vegetables and spices.

Treated food will carry a green label (see illustration) with the RADURA, the internationally recognized symbol for use on irradiated foods. Non-packaged food like loose produce should have an adjacent sign displaying the symbol. There's a catch with packaged products, however. If they contain an irradiated food that makes up less than 10% of the total ingredients (say, the chicken or mushrooms in soup), no label is required. This means that food currently imported by Canada may contain irradiated spices, undetected by Health and Welfare Canada. In theory, then, a processed food could have 11 irradiated ingredients, each making up less than 10% of the product, be 100% irradiated in other words, and not need a label. Even so, that's better than in the U.S., where labelling is required only where an entire food item is irradiated.

****

The Safety Question

There is a continuing and passionate debate over both the safety and the wisdom of irradiating food. A growing body of scientific opinion states that irradiation destroys some vitamins and other nutrients. Many scientists believe it creates entirely new chemicals and that these "radiolytic" products could be toxic or carcinogenic or both.

Although the health effects are still in doubt, there is no question that the use of the technology has great environmental implications. Critics don't like the idea of putting potentially dangerous nuclear technology into Third World countries that often suffer power cuts and where highly trained technicians and reliable maintenance are in acutely short supply. This, they say, increases the possibility of accidents causing deaths and widespread environmental contamination.

Food distributors and retailers use a lot of refrigeration equipment in their warehouses, trucks, and stores.

Miracle Foodmart and **Steinberg** are gradually replacing the air conditioners, coolers, and freezers in their stores with units that use more ozone-safe freon. The companies' Quebec division has begun filtering and reusing the freon in its refrigerated vans rather than allowing used freon to escape into the atmosphere.

Opponents make these arguments against irradiation:

■■■ It's unnecessary; present food preservation processes are perfectly adequate.

■■■ It will increase the cost of food, and that will be of no benefit to those in developing countries who can barely afford food now.

■■■ It could turn the food on our plates into poisons and cancer triggers, although we won't learn about these effects for many years.

■■■ It will create nuclear wastes and the hazards of transporting those wastes without accidents, and it will create both health and safety risks for those working and living with the technology.

■■■ It doesn't do many of the jobs its proponents say it does – long-term preservation, for example. If a food is re-exposed to bacteria after irradiation, it can become re-contaminated.

■■■ The technology is easily open to abuse. Irradiation should be used only on fresh food, but it can disguise non-freshness. There have been several instances in Europe in which seafood badly contaminated with bacteria was irradiated to make it saleable.

■■■ It encourages the nuclear industry and chemical-based agribusiness, both of which have already damaged the environment in too many other ways.

For consumers of food, the bottom line is that there is no readily available scientific test by which we or any government inspector can definitely know whether a food product has been irradiated – or at what dose. Until there is such a test, a ban on irradiated food in this country would help, even though it would be no guarantee. Yet more than 75% of Canadians are opposed to food irradiation, and both U.S. and Canadian food processors who've announced they were about to irradiate food have backed off when faced with consumer backlashes.

Those who advocate the use of irradiation say there is no conclusive case against it. Those with grave concerns say there's enough doubt and scientific evidence to more than justify delaying its implementation until we're sure the process is safe. West Germany, New Zealand, and several American states have all banned the sale of irradiated food.

****

Fruits and Vegetables

For years we've been urged to increase our consumption of fruits and vegetables for health reasons. And we certainly have. The average Canadian goes through 185 kg of fruits and vegetables a year, twice as much as Americans and an increase of about 75 kg since 1973.

Although the U.S. supplies most of the 2.3 million tonnes of fresh and processed produce we import, in all we buy from 80 other countries. Yet imported produce can be considered truly fresh only if you can afford to fly it in, preferably by Concorde. Most of ours is trucked in, and it can be weeks old by the time it appears in a store. If it has also been stockpiled by a seller because there was too much on the market at once, it may be many weeks old. (It still looks good, of course; synthetic chemicals have made sure it does.) Vitamin C is lost rapidly from many foods once they're picked, and so are other valuable nutrients.

And then there's the wax. Not the sort of wax we're used to on cucumbers and turnips, but the application of products that may bear a close resemblance to something you'd wax a floor or car with. It's not enough to be suspicious of shiny apples, peppers, eggplants, avocadoes, and tomatoes; waxes to prevent loss of moisture are routinely used on citrus fruits, melons, peaches, squashes, parsnips, and other produce as well.

The best way around these problems is to buy, whenever possible, small, locally produced, irregular fruits and vegetables, organic if possible, that have a minimum of packaging. Why should they be small? Well, the larger and more lush-looking the vegetable, the more chemicals may have been used to get it that way.

Buying local produce encourages preservation of agricultural land in your area. In addition, produce grown locally is far more likely to have most of its nutrients left, and a minimum of energy resources have been used (with fewer pollutants emitted) to transport it to you. Irregular-looking produce is produce that has almost certainly been in contact with few chemicals. It may or may not be organic, and in any case even organic products may have very slight traces of chemical residues because of spray drift and environmental contamination that farmers can do nothing about.

When you are ready to eat your produce, scrub it well in a basin of water. (Some use a mild detergent or vinegar solution before a thorough rinse, which is probably better at removing residues.) Peel the item if you're concerned about or don't know its source; blanching beforehand helps with tomatoes and peppers.

Remove outer layers of leafy vegetables like cabbage or lettuce, and flush a strong flow of water over the remaining layers. These measures won't remove residues of systemic pesticides that have been absorbed by the produce, however, and they will undoubtedly cost you the nutrients that concentrate under the skin. Like much in life, it's a trade-off.

The less you cut up vegetables, the more nutrients will be retained in cooking; for the same reason, don't soak them beforehand or add baking soda. And try not to overcook; just tender is fine. Cook them in a covered pot with a minimum of water, preferably by steaming them. Stir-frying (which can be done with broth, consommé, or a minimum of fat) and pressure-cooking also help preserve nutrients. If you have a microwave oven, see our box on that subject. Finally, never use citrus peels in cooking unless you know the fruit was grown organically; they may contain dye as well as pesticide residues.

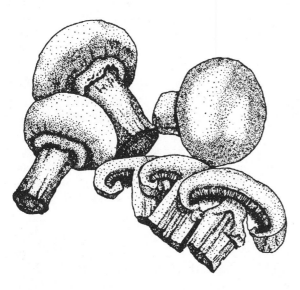

STORING FOODS

To avoid the possibility of toxic metals leaching into your food, especially acidic or fatty foods, don't cook or store anything in pots or pans that are scratched, chipped, pitted, or made of unlined copper. Use enamelled roasting pans, and don't make sherbet or fruit juice popsicles in metal ice-cube trays or store unused food in its opened can.

According to *Artist Beware*, an excellent guide to the hazards in arts and crafts by Dr. Michael McCann, "lead-glazed pottery used for eating or drinking can result in lead poisoning if the pottery is not fired properly or if the composition of the lead glaze is not adjusted properly." Although federal laws limit the lead in glazes, some pottery shipped in from the Orient can still pose a health hazard. A number of home test kits are now available for measuring the lead that may leak from decorative ceramicware.

Reserve glasses with coloured decals, logos, or slogans for display only. At the very least, don't put them in a dishwasher or let children drink from them; chances are the decals are loaded with lead. If you want souvenir glasses, stick to those with silver or gold embossing. Drinking from inexpensive plastic glasses isn't always a good idea, either.

The seal of aluminum foil can be broken by salty or acidic foods. In any case, tin foil, like waxed paper, plastic wrap, and plastic sandwich and freezer bags, isn't reusable. Use washable, reusable storage containers like old glass jars, or heavy plastic containers like Tupperware. Yogurt and margarine tubs are fine, too, but be careful when storing foods with strong flavours — once a plastic container has held chili sauce, everything you put in it afterwards will have a *soupçon* of chili sauce to it, too.

ABOUT YOUR MICROWAVE OVEN

Unlike the radiation used to irradiate food, the radiation used by microwaves is non-ionizing, the type found in light from the sun or a light-bulb, radio frequency waves, infra-red heaters, and lasers. Microwave ovens are said to be about 40% energy-efficient; that is, only 40% of the energy they use actually helps cook the food. The remainder is simply frittered away. But by comparison, electric ovens are only about 14% efficient and gas ovens 7%. So you'll save some energy costs but use a lot more electricity, if only for a much shorter time. One study found that a microwave oven saves a grand total of about $10 a year in electricity costs for the average family.

Some simple rules for your microwave:

Keep in mind that cooking vegetables in a microwave may cut down on nutrients. It depends on how many you're cooking – a small amount won't cause any great loss in nutrients because the cooking time is short and little water is used. If you're cooking large amounts, however, consider another method, such as steaming.

Use only dishes specifically made and sold for use in microwave ovens. Cover them with the glass lid from a casserole instead of any sort of food wrap. If the food should be vented, put the glass lid on crooked.

Make sure the seals around the door are kept clean. If they're cracked or broken, have them replaced immediately – and don't use the oven until they are. The radiation that can leak from microwave ovens isn't nuclear, but it isn't good for you either. Pay attention to loose hinges as well, or an oven that is bent, bashed, or skewed in any way.

Stay at least a metre away from your microwave oven when it's on, for the same reason you wouldn't put your hand against a hot lightbulb or sleep under a sun lamp. Make sure children observe this rule, too. And if you have a pacemaker, stay many metres away.

Meat and Poultry

These foods cause concern as well. In addition to the chemical residues that farm animals pick up in their feed and water, many poultry, swine, beef cattle, and dairy cows are given antibiotics. If we eat enough of this food, the next antibiotic our doctors give us may not work effectively. Livestock may also be given drugs that were never meant for human use. And Canadians allergic to some drugs, such as penicillin, have suffered dangerous reactions after eating meat.

Hormones are also administered to beef cattle to make them grow faster. Disturbingly, this practice has caused premature sexual development in children who eat a lot of beef. Use of hormones in this way is regulated in both Canada and the U.S. but enforced only by occasional spot checks. Although producers are supposed to allow for a suitable withdrawal period before slaughter, during which no new doses are given, it's clear the rules are not always followed. Britain and the other countries of the European Economic Community banned this use of hormones in 1986 and, as of 1989, allow no imports of hormone-fed meat. Trade considerations may now move Canada and the U.S. to a hormone-free point of view.

Humane concerns about the manner in which the animals we eat are raised can't be separated from these health issues. The larger the commercial operation that produces them, the more likely they are to be confined for life under conditions that we wouldn't – literally – inflict upon a dog. Disease spreads quickly in such factory-farming conditions, hence the need for antibiotics.

****

Factory Farming

Many people, when they think of farms, picture animals running in fields or strolling around barnyards. Unfortunately, many animals don't have this freedom. Increasingly, factory farming, or intensive animal agriculture, is becoming the norm.

Chickens are crowded so tightly in cages they are unable to stretch their wings. They are painfully debeaked with hot irons and kept in semi-darkness to reduce aggressiveness. Many egg farmers kill all male chicks, which are useless to egg production.

Veal calves are raised in crates where they can neither turn around nor lie down comfortably. They are taken away from their mothers a few days after birth and fed a diet low in iron to keep them anaemic and their flesh pale.

Pigs are raised in crates. They are taken from their mothers a few days after birth; their tails are removed, their ears deeply notched, and their teeth clipped — typically without anaesthetic. Sows spend most of their lives immobilized in gestation crates where they can only stand or lie down.

Cattle are often raised in crowded conditions. They are usually castrated, dehorned, and branded without anaesthetics. Dairy cows are constantly impregnated to maintain milk production.

Concern for farm animals has increasingly been emphasized by animal protection groups. Some propose vegetarian diets and others more humane farming conditions. Others focus on the enormous environmental impact of intense animal production. Intense breeding practices and mechanized slaughter use immense amounts of natural resources and energy and also generate high levels of pollution.

If you eat meat and other animal products, eat free-range and organic whenever possible, but bear in mind that free-range does not necessarily mean organic. More stores, including supermarket chains, are carrying organic meat and free-range eggs. If you can't find what you want, ask your grocer why these products aren't stocked.

* *

Meat and the Environment

According to John Robbins, author of *Diet for a New America* and founder of EarthSave, meat production is inextricably linked to environmental destruction. Here are some of the realities about the ecological effects of a meat-centred diet that he presents:

- To produce 1 kg of edible beef requires 16 kg of grain and soybeans
- 1 hectare of prime land can yield 7,350 kg of potatoes or 46 kg of beef
- 85% of topsoil lost in the U.S. is directly related to livestock raising
- To produce 1 kg of edible beef requires 43500 L of water; to produce the same amount of wheat requires 200 L
- 50 times more fossil fuels are needed to produce a meat-centred diet than a vegetarian diet
- For every person who switches to a meatless diet, half a hectare of trees is spared every year

Obviously, the food choices we make can have a tremendous effect on the environment. In fact, it is one of the strongest tools that people have to make a difference to the health of the planet. So, what can you do? Eat low on the food chain whenever possible. You eat high on the chain when you eat animal products (meat, eggs, milk) and low on the chain when you eat plants (fruit, vegetables, and grains). This is not only good for the environment — it's also good for you. One reason is that toxic chemicals such as pesticides tend to build up as you move up the food chain.

You don't have to become a vegetarian overnight. Just gradually give up meats and other animal products and replace them with grains, legumes, fruits, and vegetables. Try to make a habit of eating a meatless meal once or twice a week. You'll be doing both yourself and the environment a big favour.

* *

Fish and Fish Farming

Aquaculture now produces 12% of the world's total fish harvest. The family-run "farms" previously typical of the business have become an industry that is the equivalent of agribusiness. On the face of it fish farming sounds terrific. Basically, it involves raising fish in enclosed ponds or pens, most often pens built in natural waterways. The major environmental problem with fish farms is all the waste that goes into the water. Fecal manure and uneaten feeds form a blanket under the fish pens that smothers the habitat of wild fish and uses up oxygen and other nutrients as they decompose, which in turn stimulates the growth of weeds and algae. Also in the waste matter are cleaning chemicals, herbicides used to control aquatic weeds, and drugs used to treat diseases and parasites, as well as soluble products like the ammonia that fish produce through metabolism and any parasites and bacteria the hatchery may harbour.

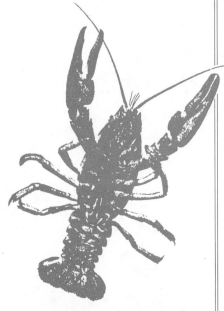

Regulatory authorities say growth hormones are not used on fish farms in Canada. Antibiotics are administered only when disease breaks out, and tranquillizers when big fish have to be handled to milk their sperm or extract eggs. At present, the federal government allows no antibiotic residues in any fish sold — although the technology for testing fish for such residues won't be available for several years.

There is serious concern that salmon from farms may spread new diseases to wild stocks if they escape from their farm pens, and that they will reduce the variety in gene pools.

Finally, there's the question of dyes. Although many B.C. salmon farmers do use dyes, the amount is unlikely to pose a serious health hazard. However, only strong consumer demand is likely to bring clear data from our governments, as well as a tough and well-organized code of environmental safeguards at every Canadian fish farm.

Until fish are identified as "farmed" or "wild" in the marketplace, you'll have to follow your own guide: from September to May, any fresh Canadian salmon or trout you buy is almost certainly from a farm. Shellfish is usually identified because producers believe consumers prefer "cultivated" oysters and mussels.

Fleets fishing for tuna in the southeastern Pacific have put dolphins in great danger. Swept up in the nets with the tuna, they are dying at the rate of up to 125,000 a year.

What about the non-farmed fish and seafood you buy? Overfishing puts so much stress on many ocean fish species that their average weights fall. Big, technologically advanced factory trawlers vacuum up the small with the large. Our lakes, streams, and coastal waters are affected by fertilizer and pesticide runoff; industries still dump their wastes, legally and illegally. The sale of trout from Lake Ontario is banned, yet fish from any waters may now contain residues and industrial toxins. PCBs, banned for years, still show up in significant concentrations; so do dioxin and chlordane.

This depressing litany does not mean you can't eat fish. Fatty fish such as carp, catfish, white perch, and mackerel tend to have the highest concentrations of contaminants; offshore species the least. Cod, haddock, flounder, pollock, salmon, and shrimp are likely to have little.

If you like fish and want to continue eating it, follow these suggestions for minimal exposure to possible contaminants:

▬ Learn what species are likely to have high residues and eat them only once a month.

▬ Ask where the fish you buy come from; choose something else if it originated in waters close to a major city or industrial area.

▬ Cook your fish only by broiling, baking, or poaching, and don't make sauces from the drippings or poaching water.

▬ Trim away the fattiest tissue (the skin, belly flap, and dark meat) after cooking.

▬ Eat little fish that comes in undercooked or raw form (ceviche, sashimi, sushi) and don't eat raw shellfish.

▬ Buy fish canned in water rather than oil, and rinse it off before serving.

****

Dolphins and Drift Nets

Finally, consider the dolphins. Greenpeace has started a grass-roots campaign to boycott certain varieties of tuna because fleets fishing for them in the southeastern Pacific have put dolphins in great danger. Swept up in the nets with the tuna, they and other marine mammals and birds are dying at the rate of up to 125,000 a year.

Responding to consumer boycotts, the world's three largest producers of tuna — Starkist, Chicken of the Sea, and Bumble Bee — have promised not to purchase tuna from fleets that have caused the deaths of marine mammals. Legislation is pending in the United States to require "dolphin safe" logos on packages. To ensure that we do not become a dumping ground for tuna that is not dolphin-safe, Canada needs to pass similar laws.

A resolution, supported by Canada, to end the use of drift nets is currently under consideration by the United Nations. Canada and other nations have outlawed the use of drift nets within their territorial waters.

From September to May, any fresh Canadian salmon or trout you buy is almost certainly from a farm.

Dioxins in Paper Products

Dioxins, a chemical family with 75 members, are some of the most lethal poisons ever created. Even small amounts can trigger a wide range of health effects, including suppression of the immune system and birth defects. The growing dioxin contamination of the environment has concerned North American scientists and researchers since 1980. Found in the effluents from papermaking processes, they contaminate rivers and lakes; garbage incinerators also release dioxins and the closely related furans into the air, to fall on water and soil.

Many paper companies are changing their processes, but dioxins are still pervasive. They are in all bleached paper products, such as paper towels, writing paper, and paper napkins. Although Canadian law bans any dioxin residues in food at all, they recently showed up in cardboard milk cartons – and in the milk inside them. Residues are also commonly found at low levels in fruit, meat, eggs, and vegetables, and have even appeared in breast milk. Buy unbleached paper products whenever you have the opportunity; demand created by Green Consumers could make a difference to the paper industry.

****

Grains

It's possible to find packaged bread free of preservatives or other additives, but these would have been added during processing, not growing: the grains in the field may still have been sprayed with pesticides. For breads made from organically grown grains, check a health-food store.

And don't assume that the darker the bread, the better it is for you. Questionable colourings may have been added to make it look more "whole grain." As stated in *Additive Alert*, "You can be sure that any rye bread that looks like chocolate cake has added colour!"

The tidbits and treats of life – cakes, cookies, muffins, biscuits, and breakfast cereals – can be a double-barrelled threat. As well as containing residues and additives, they're likely to be made with the heavily saturated tropical vegetable oils (coconut, palm, or palm kernel), butter, or even beef fat. If you want to check the fat content of a biscuit or cookie, rub it on a paper napkin; if it leaves a grease spot, you can believe it has at least 50% fat.

As with everything else, check the labels. Look for the type of fat used ("vegetable oil" isn't enough information if you're trying to avoid saturated fats) and how high it comes on the list of ingredients. If you're watching cavities as well as calories, also note how high sugar or any word ending in -ose (sucrose, glucose, fructose, and such) appears in the list. Even organic breads, cereals, and goodies may contain unhealthy fats or sweeteners.

****

Eggs and Dairy Products

Many of the cautions that apply to other foods from animals should be considered when you go down the dairy aisle. With these products too you can meet the problem of antibiotics and hormones because, of course, eggs, butter, milk, cheese, yogurt, and such come from animals that may have been given drugs. And a study conducted for the Toronto Department of Public Health found the heaviest concentration of pesticides and dioxins in dairy products.

Organic dairy products and free-range eggs are available as substitutes in many areas. But it is next to impossible to buy organic fluid milk. And free-range eggs are not regulated and are not always organically produced. If non-organic products have to be your choice, you'd be wiser to reach for skim milk, sour creams, and cottage cheeses, since the higher-fat products have higher levels of chemical residues.

When buying eggs, choose cardboard containers over spongey foam plastic. Buy milk in refillable glass or plastic bottles in the few communities where they are available. If your local dairies still use cardboard containers, encourage them to switch.

The plastic containers that yogurt, margarine, and other dairy products come in are not environment-friendly when you discard them, but at least you can reuse them many times before you throw them out.

Fats and Oils

You'll be glad to know that any pesticide residues that may be in the vegetable sources used to make oils for cooking and salads are reduced during commercial processing to undetectable limits. And there are organic cooking oils to be found, as well.

No vegetable oils contain cholesterol, which comes only from animal sources, but some oils are far more saturated than others, a point that should be taken into account if you're one of the many Canadians trying to lower their consumption of fat in general and saturated fat in particular. As you've learned, there is now an additional but vital reason for doing so: toxic residues moving up the food chain are retained in the fat we get from meat, animal products, fowl, and fish. In short, these fats can give us much more than cholesterol.

Note, however, that the fat restrictions adults may place on themselves should not be extended to children without consultation with a doctor or nutritionist. The young need fats for growth and development; without a sufficient amount, they can become malnourished.

* * * * * * * * * * * * * * * * * *

Hot and Cold Beverages

To the environmentalist, the first question about beverages is: what sort of container do they come in? Buy cold ones in returnable or recyclable containers whenever possible. With hot beverages, try to avoid the foam plastic containers into which they are forever poured; carry your own plastic mug with you and get the office to purchase a set of washable mugs.

Coffee and Tea

Cultivation of land for coffee, our favourite hot drink, has led to the wholesale clearance of forest in many parts of the world, and to the exhaustion of fragile soils. Coffee crops, being prime examples of the monoculture system of growing, are sprayed with large quantities of pesticides, some of which may be illegal in North America. As well, coffee-washing plants pollute many rivers with very strong effluents, and enormous amounts of energy are used to roast, grind, and process coffee.

Organic coffees are becoming easier to find, although finding teas is still a bit of a challenge. If you drink decaffeinated coffee, look for a brand that is produced by a water process rather than a chemical one. And remember that bleached coffee filters contain dioxins, produced during the heavily polluting paper-bleaching process. Use unbleached cotton reusable filters or reusable metal cones.

* * * * * * * * * * * * * * * * * *

Fruit Juices

First make sure it's really fruit juice, and not just a fruit drink. Then try to buy it in large bottles or cans instead of in non-recyclable drink boxes. There are few organic juices, but it should be fairly easy to find apple cider that's organically produced. Keep in mind that although Alar has been withdrawn from the Canadian market, it stays in the fruit of apple trees for three years after their last spraying.

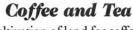

ORGANIC Wines

These wines are additive-free and made from grapes grown by organic viticulture.

FRANCE
Champagne and Méthode Champenoise:
Carte d'Or Champagne José Ardinat
Saumur Méthode Champenoise Brut Gérard Leroux
(Red Wines:)
(Midi, Provence, and South)
Domaine de Clairac Jougla Vin de Table
Domaine de l'Ile, Vin de Pays de l'Aude
(Bordeaux)
Château du Moulin de Peyronin
Château Renaissance
Château de Prade Bordeaux Supérieur
Château Méric Graves
Château Barrail des Graves St Emilion
Domaine St Anne Entre-deux-mers
(Rhône)
Cave la Vigneronne Villedieu
Vignoble de la Jasse
(Burgundy)
Macôn Alain Guillot
Bourgogne Alain Guillot
(Beaujolais)
Château de Boisfranc Beaujolais Supérieur
(White Wines)
(Loire)
Blancs de Blancs Guy Bossard
Gros Plant du Pays Nantais sur Lie Guy Bossard
Muscadet de Sèvre et Maine sur Lie Guy Bossard
Sancerre Christian et Nicole Dauny
(South of France)
Mauzac Vin de Pays de l'Aude
Limoux, Domaine de Clairac
Chardonnay Vin de Pays de l'Aude
Pétillant de Raisin
Coteaux des Baux-de-Provence Terres Blanches
(Bordeaux)
Château Ballue Mondon Sec
Château Ballue Mondon Moelleux
Château Meric Graves Supérieur
Château le Barradis Monbazillac
(Burgundy)
Bourgogne Rouge Alain Guillot
(Alsace)
Sylvaner Pierre Frick
Klevner Cuvée Spéciale Pierre Frick
Gewurztraminer Pierre Frick
(Rosé Wines)
Rosé d'Anjou Gérard Leroux
Domaine de Clairac Jubio Rose

SPAIN
(Red Wines)
Biovin Valdepenas

ITALY
(Red Wines)
Chianti Roberto Drighi
Valpolicella Classico Superiore
(White Wines)
San Vito Verdiglio Roberto Drighi
San Vito Bianco Toscano Roberto Drighi
Soave Classico Guerrieri-Rizzardi

ENGLAND
Organic Apple Wine – Avalon Vineyard

For a fee, CPRT Lab Inc. will test your tap water for 70 different substances, including PCBs, pesticides, and sundry carcinogens. For information, send a stamped, self-addressed business-sized envelope to: CPRT Lab Inc., 22 Gurdwana Road, Nepean, ON K2E 8A8.

Bottled Waters

The state of our drinking water is a pressing problem that Green Consumers should tackle at the political level. According to *The Supermarket Tour,* by the Ontario Public Interest Research Group, bottled water is the fastest growing segment of the $300-billion worldwide beverage market. Largely because of a growing concern for the safety of our tap water, Canadians drink an average of four litres a year.

Buying bottled water, however, does nothing to remedy the pollution of the water that continues to flow from our taps. Moreover, most bottled waters are no better than the tap water their buyers are trying to avoid. A study conducted by the Consumer's Association of Canada tested 15 brands of bottled spring water and tap water from seven Canadian cities. All of the bottled waters contained dissolved metals, inorganic chemicals, and other contaminants, often at higher levels than tap water. According to the Canadian Public Health Association, consumers are at greater risk from bottled water because tap water is more strictly monitored.

Demand that governments provide clean tap water rather than making it a luxury available only to those who can afford to pay for it. If you are concerned about the water quality in your area, have it tested by your municipality or province. If they determine it is below standard, they will be able to advise you on exactly what to do about it.

****

Beer, Wine, and Liquor

The scotch whisky industry's need for peat has led to conflicts with conservationists because peat extraction on the island of Islay is threatening to destroy an important feeding ground for migrating geese. And the fermentation processes involved in producing liquors like gin and whisky have traditionally produced very strong effluents that caused recurrent water pollution problems. In recent years, however, the industry has invested heavily in effluent treatment facilities.

Wines may contain sulphite preservatives, to which some people are extremely allergic, and many a headache is caused by chemical additives rather than overindulgence. There are organic wines in Europe; see box on page 45. If they are not carried by your provincial liquor board, you may be able to import a case of 12 bottles.

West German beers contain only hops, malt, and water; no additives are allowed. Canada now has many micro-breweries that produce additive-free beer and ale. Most are affiliated with Britain's **Campaign for Real Ale** (CAMRA), and all use only malt, barley, hops, yeast, water, and, in some cases, brown sugar colouring in their beers. These are not organic beers, however. For more information, contact CAMRA at P.O. Box 2036, Station D, Ottawa, ON, K1P 5W3.

****

10 ✓ WAYS TO A GREENER DIET

1. Eat low on the food chain, adding more grains, fruits, and vegetables to your diet and reducing meat and animal products.

2. Eat organic foods, preferably Canadian ones, and don't expect a perfect appearance. Where you can't find organic, try to keep your imported food choices to a minimum; off-shore pesticide controls and inspections may be less rigid than Canada's.

3. Concentrate on in-season food grown locally; out-of-season produce is shipped a long way for a long time and is often treated with chemicals to keep it from spoiling.

4. With non-organic foods, follow our preparation and cooking instructions in each section; even with organic meat and poultry you'd be wise to remove all visible fat for dietary reasons.

5. Keep yourself informed about the pesticides used on foods and the additives used in them (see separate box on books), and ask your grocer to stock foods without them.

6. Read the labels of all processed foods, on which the ingredients must be listed in order of quantity; buy the products with the fewest ingredients listed after the food itself.

7. If you're concerned about possible chemical residues or want to complain about additives in a packaged food, write to the manufacturer listed on the label; usually only the company name, city, and postal code are listed, but that should be enough for a letter to reach "The President."

8. Consider sending a photocopy of your letter to the president of your grocery chain; the store manager can give you the name and address.

9. Consider, too, sending photocopies of your letter to your municipal, provincial, and federal political representatives, and to the departments of health at all three levels of government.

10. Take a few minutes to write a thank-you note when you are pleased about finding an additive-free, certified organic, or otherwise environment-friendly product in a store.

To Learn More

If you'd like more information about chemical residues in American fruits and vegetables, read *Pesticide Alert*, by Lawrie Mott and Karen Snyder (Sierra Club Books, 1987). For information on the environmental contaminants in Canadian food, read *The Invisible Additives*, by Linda Pim (Doubleday, 1981). And if you're concerned about the additives that manufacturers and processors contribute to your food, read *Additive Alert*, a Pollution Probe publication (Doubleday, 1991).

Food Co-ops and Buying Clubs

According to the Ontario Federation of Food Co-operatives and Clubs, food co-ops and buying clubs are two ways people can gain control over the quality, source, and cost of their food. Both exist as alternatives to supermarket shopping and function as member-owned, democratically run organizations. Food co-ops have a retail storefront usually financed through members' loans and food price mark-ups. Buying clubs are generally smaller and often group around schools, churches, and community centres. Their members pool their purchases and collectively place a bulk order.

The OFFCC is one of five member organizations of the Alliance of Co-operative Natural Foods Distributors of Canada. For information on how to start your own buying club or to become a member of a food co-operative, contact one of the organizations listed below. Most require only five adult members to start a buying club, although they do suggest that a larger membership makes it easier to meet minimum-order requirements and to save money with bulk purchases, and to continue the club if one or more members move away.

Alliance of Co-operative Natural Foods Distributors of Canada
3450 Vanness Ave.
Vancouver, BC V5R 5A9
(604) 439-7977

CRS Workers Co-op Horizon Distributors
3450 Vanness Ave.
Vancouver, BC V5R 5A9
(604) 439-7977

Ontario Federation of Food Co-operative and Clubs
22 Mowat Ave.
Toronto, ON M6K 3E8
(416) 533-7989

PSC Workers Co-operative
1-800 Viewfield Rd.
Victoria, BC V9A 4V1
(604) 386-3880

Wild West Organic Harvest Co-op
150-2471 Simpson Rd.
Richmond, BC V6X 2R2
(604) 276-2411

3. CLEANERS

THE AVERAGE CANADIAN HOME *plays host every year to about 20 aerosol sprays and another 24 non-aerosol cleansers, solvents, spot removers, deodorants, and polishes. Some experts say as much as 30% of the average family's weekly shopping bill goes to pay for cleaners.*

▲▲▲▲▲▲▲▲▲▲▲▲▲▲▲

When most people think about pollution, they picture a dirty factory with a big black smokestack and a rusty pipe spewing sludge into a nearby creek. Few of us consider the environmental dangers posed by the cleaners, disinfectants, polishes, stain removers, and all other chemical products we buy, use, and eventually throw away. Yet these hazardous chemical products can make our homes just as dangerous as most modern factories. And our household garbage can be as toxic as any pouring out of a factory sewer.

Many of us spend several hours every week up to our elbows in toxic cleaners, breathing the acrid fumes of these hazardous household products. And we do this without any proper training in their safe use, with little idea of the toxic risk, and without any protective equipment at all.

Then, when we're finished cleaning and scrubbing, we pour the toxic mess

▲▲▲▲▲▲▲▲▲▲▲▲▲▲▲

down the drain, straight into the sewer and on its way to the closest lake or river. Or we screw on the cap and toss the old can or bottle in the trash, just a garbage-truck ride away from the local dump, where the noxious residues can seep into the groundwater or leak into the air for the next hundred years.

There *is* something you can do about it. Before buying or using any household cleaner, ask yourself: "Do I really need this? Is a less dangerous alternative available?" Fortunately, homemade cleaners can be substituted for almost all commercial products in our homes, and the do-it-yourself alternatives are just as effective, always safer, and usually much cheaper.

Use a mild cleaner instead of bleach. Use hot water and vinegar instead of detergent. Use a warm water and soda solution instead of caustic oven cleaners.

Most do-it-yourself cleaners can be made from a few simple ingredients (see box), all of which are found easily in supermarkets, paint stores, or hardware stores. If you don't want to make your own, a wide range of more environment-friendly cleaning prod-

---

"I've been to day-school, too," said Alice; "You needn't be so proud as all that."

"With extras?" asked the Mock Turtle a little anxiously.

"Yes," said Alice, "we learned French and music."

"And washing?" said the Mock Turtle.

"Certainly not!" said Alice indignantly.

"Ah! Then yours wasn't a really good school," said the Mock Turtle in a tone of great relief.

"Now at ours they had at the end of the bill, 'French, music, and washing – extra.'"

"You couldn't have wanted it much," said Alice; "living at the bottom of the sea."

Lewis Carroll, Alice's Adventures in Wonderland

ucts can be found in health-food stores, in some supermarkets, and through direct-sales agents. A selection of these products is listed in The Green Directory.

****

ALONG THE SHELVES

Some of the potentially hazardous chemicals that may be found in common cleaning products:

Air Freshener: ethanol

Descalers: formic acid, phosphoric acid, sulphamic acid

Disinfectants & Bleaches: calcium hypochlorite (bleaching powder), hydrogen peroxide, phenol derivatives, pine oils, sodium hypochlorite (liquid bleach)

Drain & Toilet Cleaners: hydrochloric acid, iodine, sodium hydrogen sulphate, sodium hydroxide (lye)

General Cleaners: ammonia (or ammonium hydroxide), petroleum distillates, sodium hydroxide (lye), sodium hypochlorite

Oven Cleaners: ammonia, potassium hydroxide, sodium hydroxide

Soaps & Detergents: alcohols, various surfactants, polyether sulphates, sulphonates

Stain Removers: methanol, petroleum-derived solvents, perchloroethylene, tetrachloroethylene, trichloroethylene (these last three are primarily industrial solvents)

Window Cleaners: ammonia, isopropyl alcohol

****

Probably the most important ingredient for environment-friendly cleaning is a little elbow grease. Clean often and you won't have to rely on such strong chemical weapons in your war on dirt. And clean hard; the physical component of cleaning — the sweeping, the wiping, the mopping, the churning action in your washing machine — is often more effective than the chemical component . . . and cheaper than a health club membership!

But it doesn't have to be all hard work. The ways you clean — the special tricks you use to make the job go easier — are just as important as the chemical cleaners you use. For example, mopping up spills right away, before they set and dry, makes removing stains a lot easier.

Smart cleaning can also be better for the environment; the two-bucket method of cleaning floors and walls, for instance, makes the job go faster, gets your home cleaner, and uses a lot less cleanser.

5 S·A·F·E·R SUBSTANCES

Vinegar, baking soda, pure soap, washing soda, and cornstarch are the five foundations of environment-friendly cleaning. They are safer, cheaper, and can be just as effective as their commercial counterparts.

VINEGAR
(5% acetic acid)
Common white vinegar is derived from either petroleum products or natural grain mashes. Vinegar cleans, deodorizes and removes mildew, stains, and wax buildup. It is mildly corrosive.

BAKING SODA
(sodium bicarbonate)
The all-round champion. Baking soda cleans, deodorizes, scours, polishes, and removes stains.

PURE SOAP
Cleans everything from dishes to cars.

WASHING SODA
(sodium carbonate)
Cleans clothes and softens water. But it is moderately toxic. Wear gloves and use it in a well-ventilated area to avoid irritation of mucous membranes.

CORNSTARCH
This odourless pure vegetable powder can be used to wash windows, freshen carpets, and clean up greasy spills on countertops, driveways, boats, and garage floors.

AND 3 MORE

Borax, hydrogen peroxide, and household ammonia (ammonium hydroxide) can be used for tougher cleaning jobs. But these are more toxic compounds and should be used sparingly and carefully. Wear gloves, and use them only in a well-ventilated area.

BORAX
(sodium borate)
This odourless, poisonous crystalline powder can irritate the eyes, nose, and throat.

HYDROGEN PEROXIDE
Less dangerous than borax or chlorine-based bleaches, hydrogen peroxide can still irritate the skin, eyes, and throat. Commercial brands vary in concentration from 3% to 98% pure; mixes above 35% can easily cause blistering, which may be delayed in appearance. Take care; wear gloves and wash thoroughly if you spill any on your skin.

HOUSEHOLD AMMONIA
Cleans ovens and grills, removes floor wax, and deodorizes and disinfects appliances, tiles, and fixtures. However, contact with this strong alkaline can burn the skin, eyes, and throat. The pungent fumes can cause headache, nausea, chest pain, and sweating. **Never mix ammonia with chlorine bleach — the mixture creates a poisonous gas.**

Almost any chemical compound can be toxic — if you swallow or breathe enough of it. Green Consumers search out products or homemade alternatives that pose the least danger to their health and the health of the planet.

One bucket holds your cleaning solution while the other contains the rinse water. Dip your sponge about halfway into the bucket of diluted cleaner and spread the solution over a one-metre-square section of wall or floor. Wait a minute. Then wipe off the solution with your sponge and squeeze the dirty water into the second bucket.

A clean cloth can put a shine on the wiped area and get any residue the sponge missed. A splash of vinegar in the rinse water can also help remove soapy films from floors and walls.

Your bucket of cleaner should last through the whole job. You won't have to keep emptying and refilling it with fresh cleaner, because it won't ever get too dirty to use.

✳✳✳✳✳✳✳✳✳✳✳✳✳✳✳✳✳✳✳✳✳✳✳✳✳✳✳✳✳✳✳✳✳✳✳✳✳

Four Steps Towards Cleaning the Smart Way

1 Eliminate loose surface dirt. Sweep, dust, or vacuum away dust and cobwebs before you pull out the liquid cleaners and get down to the heavy work.

2 Apply enough cleaning solution to do the job properly. You may have to experiment a little to find what quantity works best, but don't be too stingy.

3 Then let it sit. Don't be too eager to wipe it up right away. It took a long time to get this dirty. It'll take a while for your cleaner to dissolve away the dirt.

4 Finally, remove the mushy dissolved dirt. Use a sponge, clean rag, or squeegee to finish the job with ease.

✳✳✳✳✳✳✳✳✳✳✳✳✳✳✳✳✳✳✳✳✳✳✳✳✳✳✳✳✳✳✳✳✳✳✳✳✳

Aerosols

Don't buy cleaning products (or other consumer products, for that matter) that come packaged in aerosol cans. Whenever available, use pump-spray bottles or other containers instead. Even though the manufacturers have taken the ozone-destroying CFCs out of aerosol products, the containers are still non-recyclable (as yet in Canada, although pilot tests are under way in other countries) and a waste of resources when thrown away. Aerosol cans are also a possible explosion or fire hazard when heated or punctured. Discarded aerosol cans are liable to blow up when crushed in the back of a garbage truck or in an incinerator or recycling plant. For proper disposal of aerosols and other household hazardous waste, see Chapter 7.

They can pose a health threat. An aerosol product has three main components: the active ingredient(s), the propellant, and the various additives (such as plasticizers, resins, surfactants, emulsions, and so on). Aerosol propellants currently in use in Canada include flammable hydrocarbons (butane, propane, and isobutane), nitrous oxide, methylene chloride (a suspected human carcinogen), carbon dioxide, dimethyl ether, and nitrogen. Some of these substances may present serious health hazards. To make matters worse, aerosol cans deliver their contents in a fine mist, which you will inhale. This may result in the irritation of your nasal passages and upper respiratory tract and/or the speedy absorption of any toxins into your bloodstream (and hence the rest of your body) via your lungs.

Air Fresheners

Air fresheners," both sprays and solids, are frivolous at best and dangerous at worst. Most simply mask the unwanted odour with a much stronger one. Some industrial air fresheners use chemicals that actually numb the nerves you use to smell. They may also contain a wide variety of chemical substances (as well as the propellants in aerosol products) that you inhale along with the fragrance.

There are simpler and safer alternatives. Start by keeping things clean. And then guarantee good ventilation in your bathroom, kitchen, and bedrooms — simply open a window or use an exhaust fan when odours build up. Designate a well-ventilated room in your home as the only room where smokers can light up — or better yet, ban smoking altogether. Fresh-cut flowers can brighten and freshen any room. A dish of homemade potpourri is harmless, attractive, and chic, and can be bought in bulk rather than in wasteful plastic packaging. Or wrap some cloves and a cinnamon stick in cheesecloth, put it in a pot of boiling water, and let it simmer on the stove for a while. A lighted beeswax candle can burn gaseous odours away. And a bowl of vinegar will absorb strong odours and tobacco fumes.

❊❊❊❊❊❊❊❊❊❊❊❊❊❊❊❊❊❊❊❊❊❊❊❊❊❊❊❊❊❊❊❊❊❊❊❊❊

All-Purpose Cleaners

Commercial liquids used for floors and general household cleaning contain varying combinations of toxic materials, with higher concentrations found in the "heavy-duty" versions. The following all-purpose cleaners will provide a safer, low-cost alternative.

ALL-PURPOSE CLEANER

50 mL borax
125 mL pure soap
1 drop eucalyptus oil
4 L hot water

Mix thoroughly. Cleans bathroom fixtures, countertops, floors, tiles, and painted walls. After washing, rinse with clean water. For regular weekly floor washings, use a mix of mild soap and warm water. Remember, a splash of vinegar in the rinse water helps remove the soapy film.

WALL & FLOOR TILE CLEANER

125 mL shredded coarse soap
125 mL washing soda
4 L hot water

Dissolve ingredients in hot water. Using a stiff brush, scrub the tiles with the cleaner. Rinse well (soap-based cleaners tend to leave a film) and dry.

Another simple cleaning solution can be made by mixing 50 mL washing soda in 1 L warm water. Wash, then rinse with clean water.

Many guides recommend mixtures of baking soda and vinegar (with perhaps a dash of borax or ammonia). This is useless as a cleaner — the baking soda and vinegar neutralize each other in a short, fizzy reaction that leaves you with a bucket of salt water.

Greening the Laundry Room

▬ Don't do your laundry until you have enough for a full load.

▬ Use soap instead of detergent (unless you live in a hard water area).

▬ Presoak stained or very dirty clothes, overnight if necessary.

▬ If you do use detergents, experiment with the amounts suggested by the manufacturer. These recommendations are based on average water conditions — in soft-water areas you won't need as much, while in hard-water regions you'll need to add more cleaner. Contact your municipal public works department for information on local water hardness.

▬ Rinse in cold water (the rinse cycle is a dilution process that is just as effective in cold water).

▬ Hang dry your clothes. If you do use a dryer, add a small wet towel at the end of the cycle to reduce static cling.

Bleach

Bleach is mildly corrosive and must be kept out of the reach of children. Use commercial bleaches sparingly; although the contents of various brands differ, the by-products of chlorine-based bleaches may contribute to water pollution. In many instances, you can avoid using bleach altogether by soaking particularly dirty items in cool water before laundering.

As an alternative to chlorine bleach, or for laundering delicate fabrics, try hydrogen peroxide–based bleach products. Soak garments in mild (less than 15% hydrogen peroxide) solutions, then rinse.

Many bleach substitutes cannot be labelled "disinfectants" as they do not conform to federal standards for sanitizing products. Vinegar, for instance, can be used for bleaching, but it is not a disinfectant. Hydrogen peroxide disinfects, but only in very strong concentrations.

* *

Descalers

Removing the scale that accumulates in your kettle, steam iron, humidifier filter, showerhead, and drip coffee maker will increase their energy efficiency by helping them to work faster. There are commercial products for descaling appliances, but vinegar and water will do the trick more cheaply.

Use a solution of one part white vinegar, two parts water. Pour some into your electric kettle and let it boil; rinse the kettle thoroughly. For your iron, pour in some of the solution and let it stand for 30 minutes; rinse several times. Remove the filter from your humidifier, soak it in a pan of pure vinegar, then wash it in detergent and water.

DETERGENTS

In the 1960s and 1970s, countless television documentaries alerted consumers and governments to the dangers of phosphates in laundry and automatic dishwasher detergents. Phosphates are not harmful in themselves; they are basic fertilizing agents that generally benefit plant life. But the amounts of phosphates we were adding to the ecosystem were too much of a good thing.

An excess of nutrients can cause "algae blooms" that choke our waterways. The decaying algae use up the oxygen in the water, asphyxiating marine life. The piles of rotting plant life and the foul gases produced are characteristic of a dying lake.

In response to deteriorating water conditions, the federal government limited the allowable level of phosphates in laundry detergents, first to 20% and then, in 1973, to 5%. At the same time, a mammoth program was launched to build or upgrade municipal sewage treatment facilities. These two measures did much to lessen the threat to the environment. However, phosphates can still pose a local threat if homes are not connected to modern sewage treatment facilities.

Recently, concern has been growing about the chemicals that have been substituted for phosphates in commercially available detergents. NTA (nitrotriacetic acid), polycarbo-oxylates, and EDTA (ethylene diaminotetra-acetic acid), for instance, join with heavy metals in the water to form compounds that do not biodegrade well and are difficult for purification systems to remove. Some experts recommend that pure soaps or phosphate-containing detergents be used in homes connected to sewage treatment plants. Those houses that are not linked up to upgraded sewage works should continue to use phosphate-free products.

Phosphates, in any case, are only part of the problem. Unless the package says otherwise, your commercial detergent probably contains a variety of other chemical agents that can contribute to water pollution. Most detergents contain surfactants (which penetrate fabrics and loosen dirt), alkaline builders, sodium silicate, soil-suspending agents, whiteners or optical brighteners, bleaches, enzymes, blueing agents, foam stabilizers, corrosion inhibitors, colours, perfumes, and drying agents.

Laundry Detergents

In Canada, laundry detergents are restricted to no more than 5% phosphate content. Pure soaps, such as *Ivory*, have always been phosphate-free. They are also free of enzymes and many of the other additives common in commercial detergents. Instead, soaps contain a simple natural surfactant (made from animal and vegetable oils) that loosens dirt, softens water, and keeps the grime from resettling. Yet even "pure soaps" contain optical brighteners, preservatives, and perfumes (to mask the not-so-pleasant smell of the soap fats).

This "chemical simplicity" does not appear to impede soap's cleaning power. Tests by the Consumers' Association of Canada found soap outper-formed 26 detergents, getting clothes cleaner and brighter and removing stains. However, such tests are somewhat subjective — cleanliness is in the eye of the beholder — and you'll have to decide which laundry product works best for you.

In hard-water areas, soap tends to form an insoluble scum and leaves your clothes looking dull. Detergents don't work as well in hard water either; but they don't leave the dirty scum soap does. Water softeners, such as washing soda, must be added to your wash water if you live in a hard-water area. If you use commercial softeners, check that they are phosphate-free (there are no controls on the phosphate levels in softeners).

LAUNDRY POWDER
250 mL pure soap flakes or powder 25 to 50 mL washing soda

* * * * * * * * * * * * * * * * *

Dish Detergents

Many environment manuals suggest consumers can make their own dish-washing alternatives by mixing grated hard bar soap or soap flakes with water, heating the mix, and storing it in a capped jar. However, a number of readers have informed us this doesn't work well — it tends to leave spots and streaks on glassware (unless they are very well rinsed), and the soap congeals, making it difficult to pour from the container. Instead, try one of the phosphate-free, biodegradable dish-washing products now available. Check The Green Directory for some specific product suggestions.

* * * * * * * * * * * * * * * * *

Diapers

If you've made the switch back to cloth diapers, you may choose to sign up with a diaper service or do the laundering yourself. Here are a few tips for washing diapers at home (compiled by Absolutely Diapers!, a Toronto supplier of environment-

Don't use paper towels: they're wasteful and often bleached, using processes that contribute to water pollution.

Better

reusable sheets such as J-Cloths.

Best

real cloths that are even more reusable, such as dishcloths or rags from old clothing. Sponges are good, too.

If you hire a cleaning service, make sure the cleaning products it uses do not pollute. Or supply your own environment-friendly products — and insist that they are used.

Oven cleaners are caustic to your skin, irritating to your lungs, and damaging to the environment.

friendly cloth diapers):

Rinse: Rinse soiled diapers in the toilet (stains can be rubbed out with bar soap), then place in the diaper pail. A dry pail is simpler and more conenient, but a soaking pail reduces stains and odours, particularly if you wait more than three days between washings. Add soap, washing soda, vinegar OR hydrogen peroxide-type bleach to the water in the wet pail.

Wash: Machine or hand rinse, then wash as you would any heavily soiled load. (The hot cycle may be needed to kill bacteria.) Vinegar added to the rinse may soften diapers and reduce ammonia buildup.

Dry: In warm weather, hang your wet diapers outside on the clothesline (sunlight helps disinfect them). In winter, hang dry them indoors (and they'll help humidify your house).

If you have really hard water in your area, increase the amount of washing soda — it's a very efficient water softener.

* * * * * * * * * * * * * * * * * * * *

Disinfectants

Disinfectants kill mould, germs, and bacteria. While many of the alternative cleaners will remove dirt as effectively as commercial products and will suffice for almost all of your home cleaning jobs, they do not meet federal standards set for disinfectants. Anyway, there's no need to go overboard with powerful disinfectants that are going to be flushed out into the environment.

For special cleaning jobs, a diluted mix of 25 mL chlorine bleach in 1 L water can sanitize, kill mould, and remove stains. Remember, never mix chlorine products with ammonia or acid products, and follow all safety instructions on the bottle.

Drain Openers

You can help keep your plumbing clear and odour-free if you're careful to never pour liquid grease down the drain and always use the drain sieve. As a preventive measure, pour boiling water down the drain once or twice a week.

If problems do develop, avoid commercial drain cleaners; most of them contain corrosive sodium hydroxide. Along with oven cleaner, they are the most dangerous products in your cleaning cupboard. And if they back up into your sink, they can permanently damage the ceramic surface.

Try using a plunger first. If that doesn't suffice, this safe substitute should.

NON-CAUSTIC DRAIN OPENER

125 mL baking soda
50 mL white vinegar
1 kettle boiling water

First pour the baking soda, followed by the vinegar, down the drain and leave for 15 minutes. When the fizzing stops, pour in the boiling water.

As a last resort, try using a plumber's snake (available at any hardware store) or call the plumber. If one drain cleaner doesn't work, *don't add a second brand*. Many of these products, when mixed, produce poisonous or corrosive fumes. Always read the label before using any commercial drain opener or declogger.

DRY CLEANING

The solvents used by dry cleaners are hazardous to the environment, especially if they find their way into groundwater. About 80% of operators use perchloroethylene, sometimes known as "perc," which is a hazardous and potentially carcinogenic chemical. Most of the others use Stoddart solvent, a petroleum distillate, which is often released directly into the atmosphere and is known to contribute to the formation of low-level smog. Some cleaners use a fluorocarbon called F-113, which harms the earth's ozone layer.

Most dry cleaners filter, distill, and reuse their cleaning solvents. The technology is available for dry cleaners to retrieve and reuse an extremely high percentage of their perc. Operators who do so find their costs reduced dramatically. There are also service companies in some areas that will collect distillation sludge and filters loaded with perc. This is becoming more common, but not all dry cleaners take advantage of these recycling options, and a lot of solvent ends up being released into the air or discarded in landfills. There, the perc can dissolve toxic chemicals out of the other garbage and move them into the groundwater and soil.

Regulations and practices vary from province to province. Dry cleaners in the Atlantic provinces are furthest ahead. After a serious problem of groundwater contamination was discovered a few years ago, operators were forbidden to dump waste in landfills and must show proof of safe, legal disposal. British Columbia cleaners also seem to be practising safe disposal. Green Consumers in these provinces can rest easy when their clothes are at the cleaners; others should check with their local environment ministries about current controls on dry cleaners and the solvents they use.

The best way for you to find out what your local cleaner does with contaminated waste is to ask — and you may be pleasantly surprised. Some cleaners have certificates on their walls, showing that their waste is carted away by a licensed shipper.

Even Homemade Cleaners Can Pose a Risk

Always clearly label your homemade alternative cleaners and include a complete list of the ingredients. In case of accidents, the poison control centre will need this information right away!

Fabric Softeners

Adding 50 mL baking soda to the wash cycle or 50 mL vinegar to the rinse cycle will soften your laundry just as well as the costly commercial liquids.

As for fabric softener sheets, they are synthetics — usually rayon — soaked with chemicals not identified on the label. You can eliminate static cling without them by tossing a small wet towel into the dryer a few minutes before the end of the cycle. Remove the garments and hang them up as soon as the dryer stops, and they'll be wrinkle-free, too.

* * * * * * * * * * * * * * * *

Floor and Furniture Polishes

The hazards of commercial polishes lie in the volatile chlorinated hydrocarbons (such as mineral spirits, toluene, and other solvents) you can inhale while using them, as well as the biocides (usually formaldehyde) added to extend their shelf life. Another problem with many commercial polishes is that they seldom dry completely. So they collect dust — and you have to apply more polish more often.

Unfortunately, many homemade polishes are difficult to use, tend to go rancid, and may even cloud or darken wood surfaces. The Green Consumer favours finishes, waxes, and polishes that are free of synthetics, petrochemicals, and biocides. Contact your local environmental store (several are listed in The Green Directory) for available wood-care products.

Two other options for finishing wood or maintaining wood finishes are pure mineral oil and tung oil. Mineral oil, available at your pharmacy, should be applied sparingly, wiped on and rubbed off. The second option, 100% pure, unpolymerized tung oil (made from the nuts of tung trees), soaks into wood and hardens over time to form an attractive matt finish. (Closely fol-

low the instructions that come with the oil.) Successive coats build to a good finish. Tung oil has been approved by the U.S. Food and Drug Administration for use on salad bowls, cutting boards, and so on.

* * * * * * * * * * * * * * * * *

Metal Polishes

A clean, low-cost alternative is to soak silver items in this solution until they are clean.

SILVER CLEANER

1 L warm water
5 mL baking soda
5 mL salt
small piece of aluminum foil

Put in a new piece of foil whenever the old one turns black. Be careful when cleaning silver plate that when removing the tarnish you do not remove too much of the silver.

To polish brass, copper, or bronze, make a thick paste of salt, white vinegar, and flour. Rub it on the metal, then wash, rinse, and wipe dry. Be careful with homemade tarnish removers — strong acids can corrode and pit metal surfaces. In addition, they do not protect the surface from further tarnishing.

* * * * * * * * * * * * * * * * *

Moth Repellents

Instead of smelly (and toxic) mothballs and flakes, sprinkle cedar chips in your closets. However, if moth larvae or carpet beetles have already gotten into your clothing, you'll have to use a stronger pest control product (read the label first and follow all instructions).

* * * * * * * * * * * * * * * * *

Oven Cleaners

Oven cleaners are among the very worst of cleaning products, especially in aerosol containers: they're caustic to your skin, irritating to your lungs, and damaging to the environment. Prevention is your best bet. Use a second pan or liner to catch drips. Most of the spills you miss can be cleaned up immediately with a little baking soda on a wet cloth. If they've started to bake on, you may have to let the baking soda sit for 15 minutes for maximum results. For tougher cleaning jobs, try this safer alternative.

OVEN CLEANER

25 mL dishwashing soap
15 mL borax
1 L warm water

Mix together in a plastic spray bottle. Spray on dirty surfaces and leave for 20 minutes. You will likely need steel wool or a plastic scrubber to remove some stubborn spots.

* * * * * * * * * * * * * * * * * *
Rug Cleaners

Steam cleaning is the best way to clean carpets. If persistent stains remain, try one of the stain removal tips under "Spot Removers" or refer to one of the cleaning books recommended at the back of the book.

Dozens of potentially toxic ingredients are found in commercial rug cleaners, upholstery cleaners, and spot removers (which are similar in composition). One suggested alternative is:

CARPET SOAP

50 mL yellow soap
1 L boiling water
15 mL washing soda
50 mL ammonia

Dissolve the soap in the boiling water; add the washing soda and ammonia. Store in a well-labelled jar. Mix 30 g of the mixture in 1 L warm water and apply to the carpet with a brush or flannel cloth. Rub hard and rinse with a clean cloth and water. Dry well and air thoroughly. Do not saturate carpets; this can cause the backing to rot or mildew to form.

Odours can be removed from carpets by sprinkling them generously with a box or two of baking soda. Leave for an hour, then vacuum thoroughly.

* * * * * * * * * * * * * * * * * *
Spot Removers

These convenience products may contain a number of toxic compounds, including suspected carcinogens, such as carbon tetrachloride and perchloroethylene. It's wiser to get that shirt with the gravy stain into the wash as soon as possible. If you act quickly enough, sometimes straight water is all you need to prevent or remove a stain; always use cold water — hot water might set the stain. For spills on your carpet, sponge the rug promptly with a mixture of vinegar and water. Then sponge with clean water and pat dry.

Here are just a few of the many home-style remedies available. More are contained in the books of household tips we recommend at the back of the book. Before using any stain remover, test it on a small, hidden piece of fabric (the hem of a garment or a bit of carpet under the couch, for example).

For Blood: Sponge with cold water and pat dry with a towel. Repeat until stain is gone.

For Coffee or Chocolate: Soak in cold water, rub with soap and a diluted borax solution, rinse, and wash in hot water. Club soda can also be used to remove chocolate; rinse with clean water.

For Grease: Rub with a damp cloth dipped in borax. Or apply a paste of cornstarch and water; let it dry and brush it off.

For Ink: For an ink stain on white fabric, wet the fabric with cold water and apply a paste of lemon juice and cream of tartar. Let it sit for an hour, then wash as usual. For ink stains on carpets, do not leave the lemon juice mixture on for more than a minute.

For Perspiration: Rub with a solution of water and vinegar or lemon juice.

■ **For Red Wine:** Clean immediately with club soda. Or soak up as much wine as possible with a clean cloth, then sprinkle with an absorbent powder — such as salt, fuller's earth, powdered borax, or talcum powder. Wait till the powder becomes sticky, carefully remove, and add more powder. Repeat until most of the stain is gone, then apply a final dusting and wait two hours. Brush away, and wash with a mild detergent solution. Rinse quickly, rub dry, and air well.

■ **For Rust:** Soak with lemon juice, rub with salt, dry in direct sunlight, and wash.

■ **For Urine:** Rub with a solution of warm water and baking soda, rinse and wash.

■ **For Water:** If a broken pipe or leak has soaked an area of carpet, prop a stool or box under the wet area and ventilate well, using a fan. If the wet area is too large, take the carpet outside and hang dry.

✳ ✳ ✳ ✳ ✳ ✳ ✳ ✳ ✳ ✳ ✳ ✳ ✳ ✳
Starch

Cornstarch provides a cheap, non-aerosol alternative to spray starch.

STARCH

15 mL cornstarch
250 mL water

Combine ingredients in a pump spray bottle and shake vigorously. You can adjust the proportions to get the degree of stiffness you prefer.

Make sure the water and cornstarch is kept well mixed. Make this up as you need it — if stored, mould is likely to grow in the bottle.

✳ ✳ ✳ ✳ ✳ ✳ ✳ ✳ ✳ ✳ ✳ ✳ ✳ ✳
Toilet Bowl Cleaners

Liquid cleaners for toilet bowls often contain toxic compounds such as hydrochloric acid, iodine, and sodium hydroxide. The solid drop-in tablets and over-the-rim devices may contain sodium bisulphite and disinfectants, as well as useless dyes in many cases. A safe all-purpose cleaner (see page 43) will keep the toilet clean if used regularly. Or just sprinkle a little baking soda on your toilet brush. Vinegar can remove hard water deposits.

✳ ✳ ✳ ✳ ✳ ✳ ✳ ✳ ✳ ✳ ✳ ✳ ✳ ✳
Tub & Tile Cleaners

Foams and liquids for bathroom cleaning can be replaced by an all-purpose cleaner (see page 43).

Baking soda and a damp cloth will clean a tub as efficiently as commercial scouring powders, which may contain bleach, phosphate builders, or corrosive ingredients. You can also use washing soda or borax; clean with a rag or a plastic scrubber, rinse and dry. Use an old toothbrush to get at the grout. A little borax sprinkled on a half lemon will scour and deodorize countertops and cutting boards at the same time.

For a general-purpose scouring powder, try this recipe.

SCOURING POWDER

50 mL pure soap flakes or powder
10 mL borax
375 mL boiling water
50 mL whiting (a fine chalk powder, available from art supply or decorating stores)

Dissolve the soap and borax in the boiling water. Cool to room temperature. Add whiting, and pour into a plastic or glass container. Seal well. Shake well before using. If you want it to be more abrasive, add more whiting, 15 mL at a time, until it's right for you.

✳ ✳ ✳ ✳ ✳ ✳ ✳ ✳ ✳ ✳ ✳ ✳ ✳ ✳
Window & Glass Cleaners

In most cases, a good-quality squeegee and a bucket of warm water is all the professional window cleaner needs to do the job. Wet the window with a sponge or window wand and squeegee it crystal clean. Remember to wet the squeegee with a damp cloth before each stroke, and begin each swipe from the dry part of the window. If any spots remain, just wipe them away with your bare finger — a cloth will leave smudges.

For extra-dirty windows, a mix of one part water to one part vinegar should cut the grime. And don't wash your windows when the sun is shining; your cleaning solution will dry too fast and streak.

Environmentalists used to say you could get your windows clean with just a couple of sheets of newspaper — the ink helped remove smears. Unfortunately, times (and printing processes) have changed. Today's inks may smudge, and paper isn't as absorbent as it used to be. You'll have to experiment with your local newspaper to see if it still works to your satisfaction.

****

The Twelve Rules of Cleaning Safety

1 **Read the Label Before Using any Commercial Cleaning Product.** Every year, people die because they start using hazardous cleaning products before they've read the safety instructions on the label.

2 **Never Mix Chlorine Bleach with Ammonia or Strong Acids.** Chlorine mixed with ammonia or any strong acid (even vinegar) will form an extremely poisonous gas. Unless you know exactly what you're doing, do not mix hazardous cleaning products.

3 **Stop Using Any Chemical Product That Makes You Feel Sick.** If you start to feel dizzy, light-headed, or congested, stop whatever you are doing. Go outside and get some fresh air. Whenever you use strong chemical cleaners, leave a window open and make sure there's plenty of ventilation.

4 **Keep an Eye on Kids and Pets.** Shoo them away from buckets or open containers of cleaning solution. Babies and young children are more sensitive to chemical fumes. Keep them out of the room where you are using strong cleansers.

5 **Pregnant Women Should Take Special Care.** Some chemicals can harm the developing baby, especially during the first three months of pregnancy.

6 **Protect Your Skin.** Wear plastic gloves and check for holes in old pairs; a torn glove is worse than no glove at all, because it holds the chemical next to your skin. Wear old, loose clothes that cover your arms and legs. And change out of them when you are finished cleaning.

7 **Do Not Smoke, Eat, or Drink While Using Hazardous Cleaners.** The lit end of a cigarette can ignite flammable fumes or cause more dangerous chemicals to form in the air. If you want to eat or drink, take a break, go into a separate room — and wash your hands and face first!

8 **Don't Make Up More Cleaner Than You Need.** Most homemade cleaners work best when freshly mixed. Any leftover cleaner should be stored in a secure, unbreakable container, preferably with a child-proof cap. *A full list of ingredients should be marked clearly on the container.*

9 Store All Commercial Cleaners in Their Original Labelled Containers. And keep all cleaners in a cool, dry place, locked away from young children and pets, and separate from foodstuffs.

10 Watch Your Feet! Keep buckets out of the way, close to the wall, where people can't trip over them. Ladders and stools can also be dangerous. Use extension poles whenever possible to reach those high or out-of-the-way places. Keep water, sponges, and wet towels away from electrical outlets.

11 Be Prepared for Trouble. Keep the address and phone number of the local poison control centre next to your telephone (you can get it from the phone book, your family doctor, or the nearest hospital). And make sure everyone in your family knows what to do in a poison emergency.

12 Get Medical Help Immediately. If you have even the slightest suspicion that your child has swallowed, inhaled, or been splashed by a toxic cleanser, call for medical help immediately. Have the product container with you when you phone for help and take it with you to the hospital.

****

Isn't Anything Safe to Use?

Despite all the scare stories in the press and on television, not every chemical causes cancer or other serious health problems. If you take the proper precautions, some hazardous chemical products can be used safely. However, you must understand the chemical risks involved if you are to safeguard the health of you and your family.

Make a rule that you won't use any chemical cleaner until you fully understand all the environmental and health risks. A product label should contain, at the very least, a full list of chemical ingredients. If not, don't buy it. And make sure you follow all the safety, usage, and storage instructions on the label.

Always ask yourself, "Do I have enough information to use the product safely?" Sometimes you are going to have to do a little digging.

■■■ For instance, any good chemical dictionary (your local library should have several) will tell you about any of the hazardous ingredients found in cleaning products.

■■■ Environmental and worker-safety groups are good sources of information, particularly on the safety measures you must take and the alternatives available.

■■■ The Canadian Chemical Producers' Association has set up the National Chemical Referral Centre, a chemical information hotline for the public. This is a non-emergency service, operated during regular business hours (9:00 to 6:00 EST). Give them a call and tell them the name of the product or chemical you are concerned about. They will give you the name of an industry expert who can answer your questions. The toll-free, bilingual hotline number is 1-800-267-6666.

THINK GREEN WHEN YOU CLEAN

THE AMAZING . . . BAKING SODA!

Shines: Sprinkled directly on a damp sponge, baking soda cleans tubs, shower stalls, toilets, and tiles. Rinse well and buff.

Scours: Get rid of your scouring powder and use baking soda when you want to scrub counters, wooden cutting boards, chrome fixtures, kitchen appliances and coffee pots, cooked-on grease or oil on cookware, stainless steel sinks, plastic utensils and plates, and thermos bottles.

Cleans pots and pans: Sprinkle baking soda into a burned pot and let sit for a couple of hours; then lift the burned residue right out! Non-stick pans can be cleaned by boiling 25 mL baking soda in 250 mL water. Protect the surface with a quick wipe of cooking oil.

Deodorizes: Baking soda serves double duty: it deodorizes while it cleans!

Freshens carpets: Dust stale carpets with baking soda, wait 15 to 20 minutes, and vacuum.

Removes stains: Mix baking soda and water to form a paste. Scrub onto a stained surface, let sit for a half hour, then sponge clean.

Cleans drains: 250 mL baking soda down the drain, followed by 250 mL vinegar, helps keeps drains clean.

Note: Grocery stores are now stocking baking soda in all kinds of "green" packages. This is a marketing gimmick. It doesn't matter what brand of baking soda you buy — all baking soda is the same. So buy the cheapest.

✳✳✳

 THE AMAZING . . . VINEGAR!

Surfaces, appliances, & tiles: Full-strength vinegar will get stainless-steel surfaces sparkling clean. Dilute with water for cleaning tiles or wiping smears off refrigerators and other appliances.

Glass: Vinegar, either full-strength or diluted with warm water, does an excellent job of cleaning windows, mirrors, and other glass surfaces.

Fixtures: Bathroom fixtures come clean with vinegar. If soapy residues are too thick to wipe away easily, soak a tissue with vinegar and let it soak on the fixture before cleaning away.

Panelling: A mix of one part vinegar to two parts warm water will clean the grime and fingerprints from panelling and varnished woodwork.

The coffee maker: Clean automatic coffee maker by running white vinegar through the cycle, followed by one or two rinses with fresh water. You can save the vinegar and use it to clean the coffee maker a couple of more times.

Showerheads: Soaking showerheads overnight in vinegar will remove the lime buildup and get things flowing again.

Shower curtains: Toss your dirty shower curtains in the washing machine with a towel. Add 250 mL vinegar to the rinse cycle and tumble dry briefly.

Look for the Danger Signs

Almost all chemical products are poisonous to some degree — their danger often depends on the dosage and the way it's delivered (swallowed, inhaled, or absorbed through the skin). Unfortunately, the labels of few cleaning products contain complete information on their contents (though most briefly cover first-aid procedures). Federal regulations require that certain hazardous products be identified with special pictograms. Look for these warning signs:

CORROSIVE:
oven and drain cleaners can eat away many materials, including your skin

TOXIC:
disinfectants, pesticides, and many cleaning products are poisonous

EXPLOSIVE:
bleach, ammonia, and other common household chemicals can explode or produce deadly vapours when mixed

FLAMMABLE:
many stain removers, furniture polishes, and solvents are dangerous fire risks. *Never* use gasoline as a cleaner.

4. CLOTHING and TOILETRIES

S.Quinlan

CLOTHING PRESENTS ESPECIALLY DIFFICULT PROBLEMS

for the Green Consumer who wishes to make informed choices. Once a garment arrives on the rack of your local boutique or department store, it probably has several labels: the brand label, a label indicating the fibre content and care instructions, and maybe a union label.

▲▲▲▲▲▲▲▲▲▲▲▲▲▲▲

"Good sense is as much marked by ...a person's dress as by their conversation."

Catherine Parr Traill

But only in the rarest instances will you be able to find out by inspecting the item how the fibre was grown or spun and where the fabric was woven or knitted. These are the processes that do potential damage to the environment; some textile mills are notorious polluters.

As a rule of thumb, natural fibres – linen, silk, cotton, and wool – can be said to be preferable to artificial ones, such as acrylics. The cellulostic fibres, such as rayon, are derived from wood and other natural sources, but are heavily processed. Most artificial fibres are made from petroleum, a non-renewable resource and one that must undergo complex – and often polluting – processing before it can be used in textiles.

▲▲▲▲▲▲▲▲▲▲▲▲▲▲

Natural fibres, on the other hand, are often grown with the assistance of harmful pesticides. The cotton industry is the third-largest user of pesticides in the world. Some cotton farmers are switching to organic farming techniques, and consumers can encourage this trend by buying clothes made of organically grown, unbleached cotton as soon as they become available through Canadian retailers.

Obviously, the choices to be made about clothing are difficult ones. There is no easy way to compare the depletion of non-renewable resources with the spraying of pesticides or cruelty to animals. Still, there are ways the environment-conscious consumer can be a better clothing buyer and user.

▬ Reduce the amount you buy. Consider whether you are getting enough wear out of the clothes you already have.

▬ Repair an item or remodel it rather than throwing it away.

If it's still wearable, but not by you, give it to a friend, a relative, or to a charity rather than adding it to your garbage.

If it's too far gone, could you use it for cleaning rags and polishing cloths? That's a better practice than buying wasteful paper towels.

Shop in second-hand clothing stores. The prices are great and you'll be reducing and reusing!

* *

FUR

For the sake of human vanity, millions of animals every year are electrocuted, poisoned, and gassed on fur ranches. Millions more die an agonizing death in the jaws of a steel trap. There are many reasons not to wear fur.

Traps. Each year in Canada nearly 5 million animals are caught in leghold, snare, and conibear traps. More than 1 million of these are non-target or "trash" species (including squirrels, birds, and even pets) that are caught accidentally. The most common trap is the steel-jawed leghold, which imprisons an animal by the leg — for hours, days, weeks — until the trapper kills it. Some animals chew through their own limbs to escape from the trap. Although over 65 countries have banned the use of the leghold trap entirely, Canada has not.

Until recently, society has ignored the effect of the fur industry on the environment. The fur trade claims to provide a model for environment-friendly industry by making use of renewable resources (animals) without polluting or damaging wildlife habitat. Fur opposition groups disagree. They counter that the continual removal of millions of animals from complex ecosystems has had a serious, negative impact on the planet's wildlife. The indiscriminate nature of trapping is endangering at least three species of wildlife: the lynx, the wolverine, and the prairie long-tailed weasel.

Ranches. In Canada, just over half the fur comes from some 1,500 fur ranches, which annually execute approximately 1.4 million mink and 108,000 foxes. Ranched fur may actually involve more cruelty than trapping or snaring. Fox and mink spend their lives in cramped wire-mesh cages, suffer from injury and disease, and are subject to stress and psychological disorders. Inbreeding has resulted in serious genetic deformities. One of the most gruesome and cruel aspects of fur ranches around the world is their slaughtering techniques, which include poisoning, neck-breaking, and electrocution.

Fur ranching also involves environmental damage.

- It takes about 3 tonnes of feed to produce one mink coat and 1 tonne to produce one fox fur coat.
- Agricultural runoff from fur farms can contaminate groundwater.
- Pollution has resulted from inadequate disposal of waste, including carcasses.
- The use of poison to control birds attracted to fur farm waste has led to the deaths of birds and other animals.
- Processing the animal skins requires the use of energy and/or environmentally hazardous chemicals.
- The energy consumed to produce a ranched fur coat (7,965,800 BTUs) is 66 times greater than that required to produce a fake fur coat (120,300 BTUs).

The Green Consumer Speaks Out

The reality for the fur industry is that a majority of consumers no longer want to purchase products that promote cruelty to animals. In Canada and around the world, the fur industry is on the decline. According to Statistics Canada, between 1987 and 1989 the value of pelts fell 59%. In addition, the number of animals trapped was down 55% — a saving of more than 1.8 million animals. A nationwide survey showed that eight out of ten Canadians support the protection and not the killing of wildlife.

Furs and Native Peoples

Many Native people are afraid that a move away from wild trapping will threaten their livelihood. Ojibway native Paul Hollingsworth, founder of The Native/Animal Brotherhood, says that "fashion fur is not the native way. You don't see too many people hanging around the reserve in a fashion fur coat. Native tradition is to kill economically, causing the fewest deaths and gaining the most products from one death. Therefore, a native wouldn't dream of killing 40 little animals to create a piece of clothing one big animal would give them."

Native people have asked that their furs be separately labelled so that customers who wished to support Native people could do so. If you decide to purchase fur, don't wear ranched fur and make sure that your dollars go towards sustaining Native peoples. Support the labelling of furs trapped by Natives.

* *

The Great Diaper Debate

The most salient environmental issue in baby care products is the question of diapers: cloth or disposable? The argument is not so much *for* cloth as it is *against* disposables. For parents, understandably, the choice is not an easy one.

▬ Landfill

As many as 85% of Canada's babies wear disposable diapers. During the first two and a half years of their lives, each will contribute 7,000 or more soiled diapers to landfill. That's a total of 1.7 billion diapers, equalling 250,000 tonnes of waste every year.

Disposable diapers are not disposable; that is, they do not just go away. Some researchers estimate that disposables will take up to 500 years to decompose in landfill sites. (Cloth diapers decompose in about six months.) Even biodegradable diapers only break down into smaller pieces and do nothing to reduce the demand for landfill space or the resources required to make them.

Viruses in soiled diapers can be spread by flies, rodents, and other animals. Once in landfill sites, which are not designed to handle human waste, disposable diapers may pose health risks for sanitary workers and wildlife.

▬ Resources

Reusable cloth diapers use a fraction of the resources it takes to manufacture disposable diapers. Fewer than 10 kg of cotton is enough to supply all of the reusable cotton diapers needed by a baby during the two-and-a-half-year diapering period. Between 500 and 1,000 kg of fluff pulp and 325 kg of plastic (including packaging) are required for that same period to keep a baby in disposables. The plastic covering on disposables is a petroleum product — a non-renewable resource. When improperly disposed, this sheeting is a hazard to wildlife.

While wool takes the smallest amount of chemicals to produce, many shearing practices are inhumane. Sheep are bred to have such unnaturally thick fleece that they may die of heat exhaustion in summer and pneumonia in the winter after early shearing. Many sheep are subjected to tail-docking, castration, ear-punching, tooth grinding, and hurtful shearing methods. Almost all of these operations are performed without anaesthetic. A worker in the trade reports, "The shearing shed must be one of the worst places in the world for cruelty to animals."

Write to the Wool Bureau of Canada (820–33 Yonge St., Toronto, ON M5E 1G4) to express your concerns about sheep shearing practices. Support legislation to label clothing so that the origin of fabrics is specified and consumers will be able to support those operations whose production practices are the most humane.

Pollution

To make disposables involves bleaching wood pulp with chlorine gas to make it bright white. This produces toxic organochlorines, including dioxins and furans, which are emitted into the aquatic environment. The long-term effects of these chemicals on users and the environment remain unknown.

If Canadian families switched to cloth, it would reduce the demand on landfill sites and decrease the use of material resources such as pulp and plastics. Using less pulp would also slightly reduce air pollution from pulp mills.

Convenience and Cost

Cloth diapers match disposables in design and comfort. They are easy to put on as well as absorbent. And doctors and parents attest that babies in cloth diapers suffer less from diaper rash. This may be because cloth diapers are changed more often, and there is the option of using breathable waterproof covers or forgoing covers altogether.

Cost over two and a half years, according to *Canadian Consumer:*

1 Disposables
$2,200

2 Diaper service
$1,700
(no trips to the supermarket for disposables, no extra garbage to take out, no treks with pails of laundry).

Check the Yellow Pages or write to Alternatives in Diapering (5015–46 Street, Camrose, AB T4V 3G3) for the services in your area.

3 Wash-it-yourself
$1,070
($285 for three dozen pin-free diapers, $160 for diaper covers, $625 for laundry supplies, water, and electricity).

More money can be saved when cloth diapers are hung to dry rather than put in a dryer. They are even cheaper when they are reused for a subsequent child and, once worn out, used as lint-free rags.

With so much support for cloth diapers, makers of disposables are retaliating. And not surprisingly, since in Canada alone the disposables market is worth $400 million a year. But it's immaterial whether disposables are made thinner to take up less space in landfill and whether recycling or composting programs are set up. Disposable diapers are an example of a single-use, throwaway attitude and a waste of valuable resources.

Anti-fur and animal-welfare advocates detail practices and abuses in the fur industry that are horrifying and sickening. Not surprisingly, fur industry representatives contend that these claims are groundless or at least exaggerated. For more information on this debate, contact:

Canadian Anti-Fur Alliance
11 River St.
Toronto, ON M5A 4C2

World Society for the Protection of Animals
211 – 215 Lakeshore Blvd. E.
Toronto, ON M5A 3W9

Fur Council of Canada
120 – 1435 St. Alexander
Montreal, PQ H3A 2G4

Native/Animal Brotherhood
12 – 7 Forest Hill Dr.
Guelph, ON N1G 2E1

The manufacture of disposable diapers in Canada requires 65,500 tonnes of pulp (1.8 trees for every child), 8,800 tonnes of plastic and 9,800 tonnes of packing material — every year.

J·E·W·E·L·L·E·R·Y

W hile it may be obvious that we should avoid any jewellery made wholly or in part from endangered species, other forms of jewellery may also carry an invisible environmental "price tag."

■ Reptile skin.

Alligators, crocodiles, caimans, and lizards are now endangered because of the demand for watchbands, purses, shoes, and other accessories. The best alternatives are either livestock leather or non-leather products.

■ Ivory.

In 1989, a temporary moratorium was passed on the commercial trade in elephant ivory. Poaching still takes place, and the demand for ivory products still exists. It is being met in part by ivory traders who are encouraging the slaughter of walrus and narwhal in Alaska and the Canadian Arctic. The methods used to kill the animals are cruel and wasteful.

■ Coral.

Coral reefs are the undersea equivalent of tropical rainforests. Although they make up less than 1% of the ocean habitat, they are believed to support about a quarter of its marine life. These sensitive reefs are threatened by careless boaters, pollution, aquaria and fish collectors, scuba divers seeking souvenirs, and people buying jewellery made from coral products.

■ Precious Metals.

The processing of precious metals presents a number of major pollution problems. A great deal of cyanide is used in gold processing, for example, and this can cause widespread water pollution. Cyanide is particularly toxic to fish. Later in the process, when the sulphides are burned off, sulphur dioxide is produced, which can cause acid rain.

■ Gemstones.

The production of diamonds and other gemstones has often been associated with such environmental problems as deforestation (whether to clear the land for the mine and associated buildings, or to provide pit props and fuel), soil erosion, silt accumulation in rivers, water pollution, air pollution, and excessive water use. In some areas, like the Okavango Delta in Botswana, a planned diamond mine is threatening to disrupt the sensitive water regime of a wildlife-rich region completely.

Toiletries and Cosmetics

Walk into an outlet of one of the major drugstore chains anywhere in Canada and it is clear that the environmentalist pressures of the 1970s and 1980s *have* had an impact. For one thing, although you can still find glaring examples of over-packaging, there is less of it – for good economic reasons. But there still remain many environmental problems to be faced and positive choices to be made by the Green Consumer.

There has been growing concern about the use of animals in the safety testing of all types of products, but cosmetics have been particularly controversial because they are seen as non-essential. It is worth noting that just about every ingredient that goes into a perfume or cosmetic has had to be tested on animals. Once a substance or product is determined to be safe, it no longer has to be tested. However, every time a manufacturer makes the claim "new and improved," more animal tests have taken place. Three main types of tests are carried out:

For Toxicity: the most common test is the LD50 procedure, which aims to find the lethal dose — LD — of any given substance. A group of animals, such as mice, rats, or dogs, are force-fed with the substance until 50% die. The remaining 50% are killed and their organs analysed. Substances that may be tested include lipstick, nail polish, cologne, bleach, detergent, and drain cleaner. The LD50 test may be carried out over a couple of weeks or many years.

For Eye Irritation: the Draize eye test is the most notorious test among animal rights campaigners. Concentrated products such as shampoos or hair sprays are dripped or sprayed into the eyes of conscious rabbits (which are used because their tear ducts are structured in such a way that they cannot flush the substances away). To prevent the rabbits from shutting or clawing at their eyes, they may be immobilized in stocks and their eyes held open with metal clips. The procedure can continue for some days while scientists watch to see whether the eye is damaged.

For Skin Irritation: in one version, a patch of the animal's skin is shaved and scratched with a needle. Substances such as deodorant, soap, and cleaners are applied to the skin and held in place with a piece of gauze wrapped around the animal's body. Most animals suffer from blistered and burned skin, vomiting, convulsions, and comas. The surviving animals are killed or used in other tests.

As pressure grows from lobbying groups and consumers to abandon animal testing, industries become more interested in alternative testing methods. In fact, many large North American cosmetic companies contribute to the work of the Johns Hopkins Center for Alternatives to Animal Testing in Baltimore, Maryland, which was established with the financial support of the Cosmetic Toiletry and Fragrance Association. However, more public pressure is needed to convince large companies to fund alternative methods of research to replace live animals. Alternatives to testing on animals include:

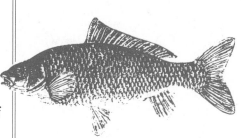

A great deal of cyanide is used in gold processing, and cyanide is particularly toxic to fish.

Marks & Spencer has not used laboratory animals in the testing of cosmetics and toiletries for more than ten years; they are the only retailer in Canada with a full line of cruelty-free products. The 77 Marks & Spencer stores across Canada carry 395 toiletries and cosmetics items. They choose only ingredients that have a proven safety record so that testing is no longer required; of course, some of those records might have been established by the use of animals in the past. Unlike most companies that follow a similar practice, Marks & Spencer also guarantees that its suppliers refrain from animal testing.

CRUELTY-FREE SHOPPER'S G·U·I·D·E

The Toronto Humane Society lists the following brands of cruelty-free skin and hair care products. A complete listing of products, including cosmetics, perfume, and household cleansers, is available free from the Toronto Humane Society, 11 River St., Toronto, ON M5A 4C2.

- Abracadabra • Aditi-Nutri-Sentials
- Aloegen • Aloette • Amway
- Annemarie Borlind • Aubrey Organics
- Autumn Harp • Aveda • Avon
- Beauty & the Beach • Beauty Counselors • Beauty Without Cruelty
- Benetton • Biokosma • Bodyline
- Body Shop • Bonne Bell • Bushwacky
- Cali • Canada's All Natural Soap
- Caryl Baker Visage • C.E. Jamieson & Co. • Chenti • Country Comfort
- Clientele • Clinique • Derma E
- Dermal Essence • Desert Essence
- Dr. Grandel • Dr. Hauschka • Enviro Green • Faces • General Nutrition Centre Products • Goldwell • Grime Eater • Gruene • Hain Pure Food Co.
- Helen Peitrulla • I+M Natural Skincare • IDI Skin & Hair Products
- Infinity • Innoxa • Isis • Jason Natural Products • Jean Nate • John Paul Mitchell Systems • Joico • Jurlique
- Kappus • Kiss My Face • KMS Inc.
- La Coupe • L'Occitane • Marks & Spencer • Mavala Nail Care • Metrin
- Micro Balanced • Mira Linder
- Molton Brown • Natural Care
- Nature Clean • Nature de France
- Nature's Gate • Naturly Rite
- Nexxus • Nutri-Metics • Ombra
- O'Naturel • Orjene Natural Products • Paul Penders • Potter & Moore • Prima Vera Aromatherapy
- Redken • Revlon • Schiff • Schwarzkopf
- Sebastian International • Shaklee
- Shikai • Soapberry Soap • Soap Works
- Sombra • Substance International
- Sunshine Fragrance Therapy • Tiki Cosmetics • Tom's of Maine • Vita-Wave Products • Webber Inc.
- Weleda • Wolf Herbal Products

■ **In-vitro Tests:** using test-tube and culture-dish studies to grow cells that can be monitored for toxicity reactions.

■ **CAM (Chorioallantoic Membrane Assay) Test:** using the membrane of the chicken egg, which contains blood vessels similar to those in the human eye.

■ **Computer Simulation and Mathematical Models:** using data collected from past experiments and research to simulate and predict reactions to chemical substances. The data can be made available to companies who wish to know the toxicity of a substance.

■ **Human Studies:** using human volunteers to test products made from ingredients already established as non-toxic.

■ **Bacteria Cultures and Protozoan Studies:** using single-celled organisms, which react similarly to human cells.

■ **Placenta:** using the human placenta normally discarded after childbirth as a perfect model of human tissue reaction to toxins.

■ **Organic Ingredients:** using ingredients already known to be safe in product formulation and that, therefore, do not need to be tested.

What Can You Do to Help?

- Buy cruelty-free. Check to see that the product has not been tested on animals, that the product's ingredients have not been tested, and that the product does not contain any animal by-products.
- Boycott animal-tested products or entire companies that conduct such tests.
- Support legislation to end animal testing.
- Make your own cruelty-free formulations. For example, many pleasant facial masks and other cosmetics can be made from natural ingredients. Oatmeal and ground almonds both make good scrubs; they help degrease oily skin. For homemade natural cosmetics recipes, look for *Cosmetics from the Earth: A Guide to Natural Beauty*, by Roy Genders.

Going cruelty-free is not an easy step for companies – product innovation may be severely stunted, production costs can be higher, reformulation can be more difficult if an ingredient becomes unavailable, and it is much harder to develop a new, exclusive substance. A company's claim that a product or ingredient has "never been tested on animals" is rarely true: usually the most that can be confirmed is that they themselves have not tested it on animals within the past, say, five years.

A growing number of suppliers try to avoid products that contain any animal ingredients, whether or not they come from endangered species. Animal products used in cosmetics include tallow, made from animal fat and used in some soaps and lipsticks; stearic acid, a solid fat found in soaps, shaving creams, and some foundation creams; collagen and gelatin, produced by boiling down bones, skin, tendons, and connective tissue; and lanolin, animal protein, placenta, and urea.

A fair number of cosmetic manufacturers still use products derived from rare species. Now that whale products like spermaceti (a white waxy substance from the head of the sperm whale, traditionally used in cosmetics) are banned in Canada, the attention of lobbying organizations has turned to creatures like the harmless, filter-feeding basking shark.

The livers of harpooned basking sharks are used to produce a refined oil called squalene, which has a low freezing point. A single six-tonne basking shark can produce 1,000 litres of oil. It is used for a range of consumer products, including cosmetic face creams. Companies using basking shark oil admit that other oils from fish and seeds could be used, but claim that such oils do not have the proven safety or efficacy of squalene. If you have concerns about this practice, write to the Canadian Cosmetic, Toiletry and Fragrance Association (24 Merton St., Toronto, ON M4S 1A1).

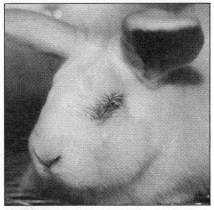

The Draize eye test is the most notorious among animal rights campaigners.

****

Aerosols

Aerosol packaging is slowly disappearing from the toiletries department. Avoid all the non-recyclable, non-refillable spray cans in favour of less wasteful forms of packaging, such as creams, sticks, and pumps.

The use of CFCs as propellants in antiperspirants and deodorants has been banned in Canada since 1980. However, they can still be used as a slurrying agent (which keeps powders in suspension in gases) in powder antiperspirant sprays. Avoid all the spray products in favour of less wasteful forms of packaging, such as creams and sticks.

****

Perfumes & Scents

Perfume production depends on animals in at least two main ways: for fixatives used to retain the scent, and, secondly, for safety testing.

Among the animal products used as fixatives are musk (taken from the musk deer), castoreum or civet (extracted from the anal sex gland of beavers or civet cats, respectively), and ambergris (from the intestines of the sperm whale). Both the musk deer and the sperm whale are endangered species. Unfortunately, because full product labelling is not required in Canada, it is generally impossible to find out what is in perfumes and other forms of scent.

The French perfume houses, which depend on many long-established recipes, are most likely to use rare, high-priced animal ingredients. Many other parfumiers have replaced such animal products, usually because of their scarcity and price rather than for humanitarian reasons. Many parfumiers have not used musk for more than 20 years, having switched to synthetics or vegetable materials.

Contraceptives

Discarded condoms can be harmful to many animals if ingested. Be careful not to litter!

****

Nail Care

Nail polish and nail polish remover contain a host of synthetic chemicals, including acetone and toluene; remover, in particular, is very toxic. If you must use these products, be sure not to let them spill into water sources, where they could cause contamination.

****

Cotton Balls & Swabs

The "cotton" in these items is actually rayon, which does not readily biodegrade. Carbon disulphide is discharged into the air during the production of rayon. Rather than these disposable products, use a washcloth.

PUMP SPRAYS ARE LESS HARMFUL THAN AEROSOL SPRAYS

Shaving Aids

If you want to take things to extremes, the greenest approach to facial hair is probably to grow a beard. For those who shave, however — more than 90% of men — the electric shaver is probably your best choice.

The amount of energy required for a typical electric shave (4.5 watts, 5 minutes per day) is only about 0.137 kWh per year. In comparison, a safety razor uses about 21.2 kWh per year just to heat the water. On top of this add the energy to make the shaving soap or cream (including packaging), the razor blades (which are disposable), and the shaver itself.

If you choose to wet-shave, avoid shaving cream in aerosol cans, disposable razors, and excessive packaging. (Particularly undesirable products are the packs of use-'em-once disposables.) Also avoid shaving brushes with badger bristles or ivory handles.

The greenest approach to shaving is probably to grow a beard.

Tampons and Sanitary Pads

Reusable tampons and sanitary pads are becoming increasingly available. The tampon or pad is simply washed out, dried, and is ready for use again.

Some disposable sanitary pads made with unbleached fibre are now available. Look for pads and tampons with the least packaging; avoid tampons with plastic applicators and pads in individual plastic pouches.

* * * * * * * * * * * * * * * * *

Tissues and Toilet Paper

Washable cloth handkerchiefs are a better choice than throwaway paper tissues. For removing makeup, use a facecloth or sponge. Look for toilet paper made from recycled, unbleached, uncoloured paper.

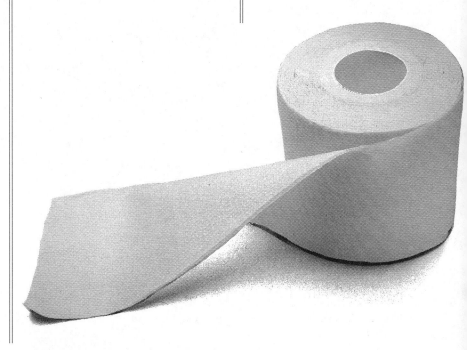

TOOTHPASTE

Toothpaste may be the last product to come to mind when you are thinking of environmental pollution, but the acid effluents produced during the manufacture of the titanium dioxide pigments used to make white toothpastes white are worth thinking about. These effluents have helped to make waterways near the manufacturing plants more acidic.

This is not a simple story of good and evil, however. Used in a wide range of paints, paper, plastics, inks, and man-made fibres, titanium dioxide has revolutionized some parts of the paint industry, where it has replaced materials like zinc oxide and lead. Unlike lead, it is completely non-toxic, and therefore a distinct improvement in terms of both health and environment. Furthermore, the two companies making titanium dioxide in Canada, N.L. Chem Canada and Tioxide Canada, are installing facilities to neutralize their acidic waste water. By 1991, white toothpaste and other products containing titanium dioxide should be back in the good books.

In the meantime, consider gel toothpastes, which contain less titanium dioxide, or natural-formula toothpastes made from ingredients such as oil of fennel, glycerine, chalk, and sodium laurel sulphate (from coconuts). And avoid toothpaste in wasteful pump containers.

✶✶✶✶✶✶✶✶✶✶✶✶✶✶✶✶✶✶✶✶✶✶✶✶✶

FOR MORE INFORMATION

Animal Alliance of Canada
1916 – 1640 Bayview Ave.,
Toronto, ON
M4G 4E9

Canadian Cosmetic, Toiletry and Fragrance Association
24 Merton St.,
Toronto, ON
M4S 1A1

Canadian Federation of Humane Societies
102 – 30 Concourse Gate,
Nepean, On
K2E 7V7

A great substitute for toothpaste is baking soda mixed with a drop or two of oil of peppermint or wintergreen. Sprinkle some baking soda into the palm of your hand, add a drop of oil, then dip a wet toothbrush into the soda. Brush as you would with toothpaste. Proper brushing eliminates the need for mouthwashes.

5. THE HOME

CANADIANS USE MORE ENERGY *per capita than any other people in the world, and they use a lot of it at home. In fact, Home Sweet Home is the third-biggest energy hog in the country (after industry and transportation), accounting for 18% of all energy consumed. And because most of this energy comes from the burning of fossil fuels, our homes are major contributors to the tonnes of carbon dioxide that get spewed into the environment every day.*

▲▲▲▲▲▲▲▲▲▲▲▲▲▲▲▲▲

"What use is a house if you haven't got a tolerable planet to put it on?"

Henry David Thoreau

The breakdown of residential energy use looks like this:

Space heating	67%
Water heating	17%
Appliances	14%
Lighting	2%

The average Canadian home gobbles about 40,000 kWh (kilowatt-hours) of energy a year. Most of it is simply wasted, primarily because of inefficient house design and construction practices, and inefficient furnaces, water heaters, and appliances.

And this isn't where the story ends. Everything inside your home, from construction materials to cake tins, accounts for more energy use in its manufacture. Many products are potential polluters, too. It all adds to the environmental load.

But the irony is that, besides being a giant energy-gobbler, Canada is also a world leader in the design of energy-efficient houses. There's the R-2000 house, for example, developed under a federal program. Annual energy consumption runs at about 25,000 kWh, just over half the figure for a conventional home. Most of the savings comes from airtight design and construction to prevent heat loss.

And then there are high-performance homes like the Ontario Advanced House, a demonstration project in Brampton, Ontario, jointly sponsored by the provincial and federal energy ministries, the Fram Building Group,

Ontario Hydro, and the Canadian Home Builders Association. It's a normal-looking 250-square-metre house, but its annual energy consumption is just above 10,500 kWh, about a quarter what a conventional house uses. What makes the difference is state-of-the-art energy-efficient technology.

Both these projects show that Canadians can drastically reduce the energy their homes consume. That doesn't mean you need to run out tomorrow and buy an R-2000 house or undertake a complete home makeover. Simple retrofit measures – like sealing leaky windows, insulating a hot-water tank, having a furnace tuned up – can reduce energy consumption by 20% to 30%. And Green Consumers can save more energy by choosing wisely when they're shopping for wallboard or wall hangings.

Individual energy-saving efforts should focus first on where it hurts most – space heating.

* *

SPACE HEATING

Two-thirds of the energy consumed in Canadian homes goes to space heating. This means that, during the winter, Canadian furnaces burning fossil fuels discharge into the environment enormous quantities of carbon dioxide and sulphur dioxide, two of the major culprits behind global warming and acid rain.

It's astonishing how much of the heat these furnaces produce simply goes to waste. It slips through cracks under doors, escapes through walls and attics, and flows up the chimney. Contrary to what many Canadians think, chilly draughts and cold feet are not inevitable features of winter in this country. The real villain here is house-construction practices developed in the heyday of cheap fuel and environmental ignorance.

There is a lot that individual energy consumers can do to spare the environment, and at the same time cut their energy bills and increase their wintertime comfort. Retrofit measures such as sealing air leaks, adding insulation, and upgrading or replacing windows can reduce by as much as 40% the energy a typical Canadian family uses to keep warm in the winter. The first step – and also the cheapest – involves caulking and weatherstripping all the spots where warm air escapes.

* *

Keeping the Heat In

If you were to add up all the draughty leaks and cracks in a typical Canadian home, you would probably end up with the equivalent of a window-sized hole in the wall. Warm air escaping through cracks and crevices around windows and

An investment of about $150 in sealing products and a do-it-yourself weekend can cut as much as 25% from the space-heating bill of a leaky house.

doors, basement walls, and other sites can account for 25% to 30% of your home's heat loss.

Sealing the air leaks in your home does involve an environmental trade-off: significant energy conservation versus small amounts of harmful by-products.

Sealants are based on synthetic chemicals; most contain solvents and produce environmentally detrimental effects during manufacture and disposal. And then there's the health issue. At least three days of good ventilation is required after the sealing job is finished to get rid of toxic vapours.

Weatherstripping involves less odious ingredients, but its use is restricted to only a few heat-loss sites – primarily the movable parts of windows and doors. Rubber weatherstripping products provide an effective and durable pressure seal; felt and foam products, on the other hand, have a poor performance record. There are two site-specific products that are well worth the few dollars they cost – electrical outlet gaskets, and door sweeps or thresholds that are attached to the bottom of exterior doors.

Rubber weatherstripping products provide an effective and durable pressure seal; felt and foam products, on the other hand, have a poor performance record.

SEALING THE LEAKS		
Air Leakage Sites	Heat Loss	Sealing Products
Basement sill plate	25%	acrylic latex, butyl rubber, polyurethane foam
Exterior electrical outlets	20%	butyl rubber, acrylic latex
Windows	13%	silicone, polyurethane foam
Pipe and wire entrances	13%	polyurethane foam
Vents (bath, dryer, kitchen)	10%	butyl rubber
Baseboards, light fixtures, electrical outlets, attic hatches	7%	silicone, acrylic latex
Exterior doors	6%	silicone, weatherstripping
Fireplaces	6%	insulated fireplace covers

If your energy-saving plans include adding insulation, you should consider installing an air-vapour barrier at the same time. It's a big job but will cut draughts and energy waste throughout the house to a bare minimum. The barrier, usually a plastic sheet, stops air leakage and keeps water vapour from penetrating into walls and ceilings. About 20 L of moist air gets deposited in a typical home every day just from routine bathing, cooking, and breathing, and it can cause structural rot as well as compact insulation materials to the point where they lose much of their effectiveness.

* *

Insulation

After caulking and weatherstripping to control the flow of warm and cold air in and out of your leaky house, the next step on your energy-conserving agenda is adding insulation to stop the direct flow of heat to the outside. The major escape routes are the walls, the roof and ceilings, and the foundation.

Before the oil crisis of the 1970s, few people worried about insulation, let alone how much they needed. Every five years or so since then, energy-efficiency experts have increased their recommended minimum levels. The following chart gives 1989 levels for new home construction.

RECOMMENDED MINIMUM RSI (R) LEVELS*

Site	Zone A	Zone B	Zone C	Zone D
Walls	RSI4.9 (R28)	RSI6.3 (R36)	RSI7.0 (R40)	RSI7.0 (R40)
Basement walls	RSI2.1 (R12)	RSI2.8 (R16)	RSI3.5 (R20)	RSI4.2 (R24)
Roof/ceiling	RSI7.0 (R40)	RSI7.0 (R40)	RSI9.2 (R52)	RSI10.6 (R60)
Floor (over crawl spaces)	RSI5.6 (R32)	RSI7.0 (R40)	RSI7.0 (R40)	RSI7.0 (R40)

*RSI (metric)and R (imperial) are measures of thermal (heat flow) resistance

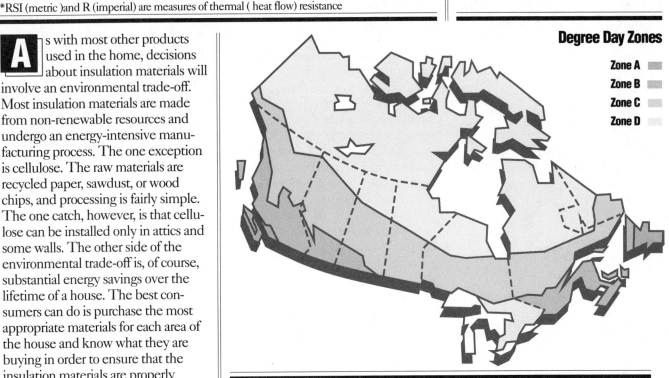

Degree Day Zones

Zone A
Zone B
Zone C
Zone D

A s with most other products used in the home, decisions about insulation materials will involve an environmental trade-off. Most insulation materials are made from non-renewable resources and undergo an energy-intensive manufacturing process. The one exception is cellulose. The raw materials are recycled paper, sawdust, or wood chips, and processing is fairly simple. The one catch, however, is that cellulose can be installed only in attics and some walls. The other side of the environmental trade-off is, of course, substantial energy savings over the lifetime of a house. The best consumers can do is purchase the most appropriate materials for each area of the house and know what they are buying in order to ensure that the insulation materials are properly installed and sealed off from the interior of the house.

Insulation materials are manufactured from three primary types of ingredients:

1 Organic polymers. The nastiest of the lot, this group includes the now-banned urea formaldehyde foam insulation (UFFI), polyurethane and polyisocyanurate foam, and polystyrene boards. These materials use an expanding agent in the foaming process: Types 1 and 2 polystyrene use air; Type 4 polystyrene (also called extruded polystyrene), polyurethane, and polyisocyanurate use CFCs. Fortunately for the environment, CFCs will be banned substances soon; *how* soon was not clear at the time of writing. For both health and environmental reasons, these products should be avoided as much as possible.

2 Inorganic materials. This group includes another banned substance – asbestos-based insulation, now found only in houses built up to the early 1970s. (It should be examined by an asbestos-disposal specialist, who can tell you whether it is better to remove it or leave it in place.) Fibrous glass generally contains formaldehyde as a binder. The better choices among the inorganic-based insulating materials are mineral wool, vermiculite (made from expanded mica ore), and perlite (made from perlite ore). Their fine fibres and dust, however, can irritate the skin, eyes, and respiratory system. Careful installation and sealing, followed by thorough ventilation, are essential.

Most insulation materials are made from non-renewable resources and undergo an energy-intensive manufacturing process.

③ Cellulose. The best of the lot, insulation products in this category include sawdust, wood shavings, and cellulose fibre made from recycled newspaper – all renewable, non-toxic ingredients. They are the friendliest choices for wall cavities, ceilings, and attics.

All insulation products use various chemicals as binders, stabilizers, and fire retardants and to control mildew and pest infestations. These can cause health problems if the insulation is not thoroughly and properly sealed off from the interior of the house.

****

Windows

In the older Canadian home, windows can account for as much as 25% of heat loss, which means that a quarter of your space-heating energy may literally be going straight out the window. You can recoup about half that loss by sealing the cracks and crevices where air rushes in and out (see pages 65-66). The other half gets lost in the flow of heat through the panes and frames of conventional windows. The only way of preventing that loss is by boosting the insulation value of your windows.

Standard double-glazed windows are very poor insulators; their R value is about one-tenth to one-sixth of the insulating value of the walls that surround them. The good news is that any effort you make to improve the R value of your windows translates directly into energy savings, not to mention increased comfort.

You can enhance the thermal resistance of double-glazed windows by installing "interior storm windows." One type is the *Window Insulator Kit* made by **3M**. It includes thin sheets of transparent polyethylene, which you cut to fit your windows. Each sheet is held in place by sticking the edges to the window trim or frame with adhesive tape, then shrinking it tight using heat from a hair dryer. The insulator adds another layer of dead air, between the glass and the plastic. A kit that will cover five 1-m by 1.6-m windows costs about $20. Most conservation stores and many of the larger hardware outlets now carry a range of these products.

Another option is movable window insulation that you open during the day to trap passive solar energy and close at night to prevent heavy nighttime heat loss. Products include window shades and blinds that are outfitted with an insulated backing material and shutters that contain a polyurethane core. For best results, they should all have a vapour barrier, a reflective surface, and a device that provides a tight seal with the window frame or trim. When shopping for these products, ensure that what you are paying for is insulation and not fancy fabrics.

Insulating values for such products as Window Warmers and Window Quilts are about R 5.5 to R 5.8. A *Window Warmer* for a 1.3-m by 1.6-m window costs about $235, not including the decorator fabric for the outer layer; for a 1-m by 1.3-m *Window Quilt*, the cost is about $340 installed. Some manufacturers offer do-it-yourself manuals and kits.

If you are renovating an older home or building a new one, consider installing high-performance windows. There has been a leaps-and-bounds advance in window technology in the past 15 years. Researchers found that adding a very thin, low-emissivity (low-E) coating of metal and metal oxide to glass drastically reduced radiant heat loss in winter. To cut down on heat lost by convection and conduction, air spaces between panes were filled with a heavier inert gas (argon or krypton); low-conduction spacers (made from

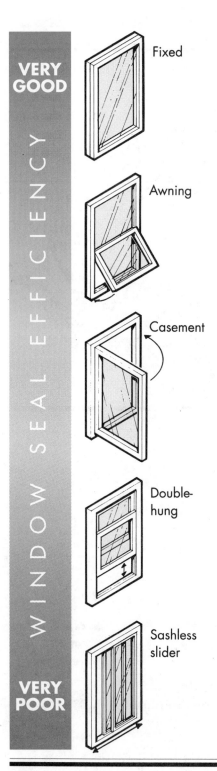

VERY GOOD

WINDOW SEAL EFFICIENCY

Fixed

Awning

Casement

Double-hung

Sashless slider

VERY POOR

polybutylene foam, fibreglass, or silicone) were inserted between the panes, replacing the highly conductive aluminum spacers used in conventional windows; a second sealant was added to guard against air leakage in the window unit; finally, frames and sashes were constructed from low-conduction materials such as wood. An added attraction of these high-performance windows is that the same technology that prevents heat loss in January also prevents heat gain in July.

The best RSI value you can expect from a top-of-the-line conventional double-glazed unit is 0.35 (R 2). With a high-performance double-glazed window, you'll get twice that, and twice again with a triple-glazed high-performance unit. Double-glazed high-performance windows cost about 12% to 15% more than conventional units; triple-glazed ones would cost 40% more than the comparable double-glazed models. Prices for high-performance windows are falling as the manufacturing technology improves, so your payback period will get shorter. Your savings in heating costs will vary according to the climate in your area, but there is no doubt that your bills will go down.

For more information about Canadian manufacturers and suppliers of high-performance windows, contact the **Insulated Glass Manufacturers' Association of Canada** (Box 1681, Brantford, ON N3T 5V7), or the **Canadian Window and Door Manufacturers' Association** (27 Goulburn Ave., Ottawa, ON K1N 8C7).

Whichever alternative you choose to upgrade the insulating level of your windows, if funds are limited, focus first on the windows that will make the most difference: those that trap solar energy (mostly south-facing windows) and those that are big losers of heat (large windows, north-facing windows, and windows in direct line with the prevailing winter wind).

****

Retrofitting an Older Heating System

A conventional heating system can be inefficient for a number of reasons: incomplete combustion when the burner is not properly tuned; poor heat transfer when the heat exchanger is dirty; heat loss through the ductwork; air flow up the chimney when the furnace isn't running. You can improve your heating system's efficiency by following a regular maintenance program, by ensuring that the heat is being properly distributed, and by upgrading or downsizing the furnace.

Maintenance. All heating systems require regular maintenance for safety and efficiency. Oil systems should be serviced every year by an oil-burner mechanic; it's about a two-hour job that involves cleaning and lubricating parts, checking safety controls, and tuning up the burner. Because gas burns cleaner than oil, gas systems need to be serviced every two years by a certified gas fitter who inspects and cleans the pilot light, the main flame, and the burner, among other things.

Improving heat distribution. There are a few small, do-it-yourself steps that will help move the heat efficiently from the burner through the ducts to where it's needed. Dirty furnace filters restrict the flow of air; they should be cleaned or replaced once a month. Check the furnace fan belt to ensure that it's tight enough to drive the fan properly. Adjust the manual dampers in the heating ducts and the heat registers in the rooms to distribute the heat where you want it. Close dampers and registers in rooms and areas of the house where heat is not required. Seal the joints in the ductwork with duct tape or silicone caulking to prevent heat loss. The ducts should be cleaned every three to five years.

INDOOR INSECT PESTS

Even if you've never sprayed insecticide in your home, you probably have some – in your new or newly cleaned carpets, for instance, which are often treated, or in wallpaper (it's in the adhesive). If you'd rather not add to the toxic vapours in your home, you might want to use safer means of dealing with common insect pests.

First try prevention:

Follow strict sanitary practices, seal cracks and other possible entry areas, and fix plumbing leaks and other areas of undue moisture.

If defence tactics fail, try natural remedies:

Cedar oil, available in manual spray pumps, repels fleas and other insects; salt or red pepper sprinkled on counters and across doors or windowsills is a barrier for ants; freezing clothing or exposing it to hot sun for two days kills moth larvae; and a vinegar-water solution wiped on kitchen counters repels flies (the old-fashioned fly swatter works, too). A borax-type bait applied in traps or drops is a safer alternative for cockroach control, but don't use it around small children or pets.

It's also worth checking out some new biological pesticides, which target specific pests without affecting higher life forms or generally toxifying the environment (they break down easily). Methoprene, for instance, is an insect growth regulator that halts flea growth at the larva stage, thereby preventing reproduction.

S. Quinton

Furnace Efficiency
AFUE: Annual Fuel Utilization
Efficiency (percent)

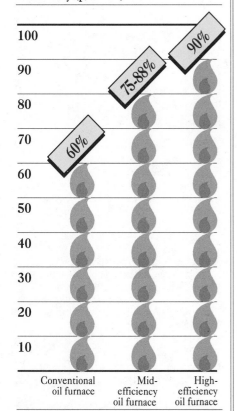

		90%
	75-88%	
60%		

100

90

80

70

60

50

40

30

20

10

| Conventional oil furnace | Mid-efficiency oil furnace | High-efficiency oil furnace |

■■■ **Upgrading.** There are a number of efficiency-enhancing devices that you can add to an older oil furnace to cut down energy consumption. Retention-head burners improve combustion and may reduce your fuel bill by 10% to 25%. A delayed-action solenoid valve ensures cleaner burning for greater efficiency, and a flue damper installed in the vent pipe prevents heat from escaping up the chimney. Consult a heating contractor before making any decisions about these devices; compatibility and the age and efficiency of your system may rule out this sort of upgrading.

■■■ **Downsizing.** If your oil burner is not running almost constantly on really cold days, you are probably wasting fuel on a furnace that produces more BTUs (British thermal units) than your home can use. Many of the furnaces installed in Canadian homes never reach their optimum level of efficiency because they are sitting idle most of the time. And when they are not running, indoor air and heat escape up the chimney. Downsizing involves reducing the furnace's firing rate or installing a smaller nozzle to restrict the flow of oil. Fuel savings can amount to 10%. To find out if these are feasible options for you, consult a heating contractor.

* *

Energy-Efficient Furnaces

I f you have a cast-iron octopus in the basement or some other furnace that has clearly seen better days, it may well be time to begin hunting for one that uses a lot less energy to produce the amount of heat you need.

The cheapest form of heating is the high-efficiency gas furnace. If you're not in an area served by natural gas, electric furnaces are very clean, reliable, and inexpensive to install and repair. An add-on heat pump will also allow you to cut your energy bills a bit while providing an efficient air-conditioning system.

How efficiently a furnace performs is measured by its AFUE (annual fuel utilization efficiency). For example, the AFUE of a conventional oil furnace is about 60%; the rating for a new mid-efficiency furnace ranges between 75% and 88%; and the AFUE of high-efficiency oil furnace is over 90%. These much higher ratings are partly attributable to what's known in the furnace trade as "efficiency-enhancing equipment" – things like flame-retention burners, delayed-action solenoid valves, automatic flue dampers, and condensing units. Other advanced features include ceramic or stainless steel combustion chambers and new, improved heat exchangers.

* *

Integrated Systems

I f both your furnace and your water-heater are on their last legs, consider replacing them with an integrated space- and water-heating system. The integrated system has one heating unit that performs the work of two conventional units, with a lot less hardware – a big energy-saving advance for energy-efficient houses with relatively low space-heating requirements.

The *Habitair*, made by **Fibreglas Canada,** is an integrated system suitable for a 170- to 280-square-metre home with good insulation. The cost is $4,500 installed.

Thermostats

By now, most consumers have acquired the habit of turning down their thermostats at bedtime and when no one is home. Lowering the thermostat just 5 °C during these periods can add up to a 14% reduction in your energy consumption.

A cost-effective investment for both old and new heating systems is a programmable device called a thermostat setback. With up to four setback adjustments possible in a 24-hour period, it allows you to automatically control the supply of heat – to keep your home at the lowest comfortable temperature throughout the day. There are a wide variety of automatic setbacks for each type of heating system; the more sophisticated ones offer weekday and weekend programming, weather monitoring, power reserves, and setback modules for a hot-water heater and air conditioner.

The location of your space-heating thermostat or automatic setback is important. It should be installed where it is not affected by direct sunlight, radiators, hot-air ducts, appliances, outside doors, or stairwells. The best place is on an inside partition wall where it can accurately register the temperature of the house.

Suggested Thermostat Settings

21°C when relaxing
20°C when working or exercising
18°C when sleeping
16°C when no one is home

Heat Pumps

Because their energy source is clean, free, and inexhaustible, heat pumps are an exciting alternative to conventional heating and cooling systems.

Air-to-air heat pumps, the most commonly purchased variety, work best in the fall and spring; in most areas of the country you will probably still need a furnace for the coldest parts of winter. In warm weather the pump extracts heat from the inside and pumps it outside; in cool weather it extracts heat from the outside air and pulls it inside.

Ground-source heat pumps extract what's called "low-grade heat" (between -2° C and 10° C) from a few feet below the ground by passing a refrigerant gas through a compressor and two heat exchangers. In the first exchanger, the gas evaporates as it absorbs ground heat, and after moving through a compressor where its temperature is raised, it passes through the second exchanger where the heat is removed and pumped into the space-heating ducts. In the summer, the system is reversed to draw hot air out of the home. With the addition of another heat exchanger, this extracted hot air can be used for water heating. A groundwater heat pump operates on the same principle but extracts heat from underground aquifers.

If you are renovating an old house down to the bare bones or building a new one, installing a heat pump is worth considering, especially if you need a cooling system as well as a heating system. Annual energy savings with a heat pump in your home can be as high as 65%.

Lowering the thermostat just 5°C during certain periods can add up to a 14% reduction in your energy consumption.

ANNUAL ENERGY SAVINGS WITH A HEAT PUMP CAN BE AS HIGH AS 65%

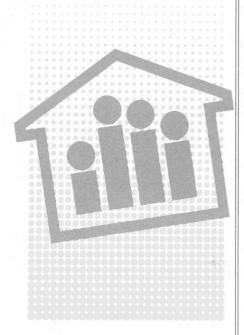

Air-to-air heat pumps cost $2,000 to $3,000 installed. Ground-source heat pumps range from $8,000 to $12,000 installed. **The Canadian Earth Energy Association** (228 Barlow Crescent, Dunrobin, ON, K0A 1T0) will answer all your questions about ground-source heat pumps and provide a list of suppliers.

* *

Ventilation

Ventilation isn't a problem if air leaks and draughts are common features of your housescape. Air moves very quickly and efficiently in and out of a leaky house; there's a complete change every 30 minutes to two hours.

To conserve energy, your house has to be as airtight as possible. But sealing all the sites where heat and cold exit and enter means that fresh air can't get in and stale air can't get out.

Consumers who want an accurate measurement of their home's airtightness should ask a heating or ventilation contractor to do a fan depressurization test (the cost is about $250). If the test results show less than one-half to one-third air changes per hour (all the air in the home is replaced every two to three hours), the home needs a mechanical ventilation system that controls the flow of air in and out of the house.

If the ventilation problem isn't too acute, all that may be required is an intermittent running of kitchen and bathroom fans. Opening a window or two for short periods of time is not a good solution; it defeats all air-tightening efforts to conserve energy without necessarily providing adequate ventilation.

Designers of energy-efficient homes have solved the air-quality and ventilation problems by installing heat recovery ventilators (HRVs). These air-to-air heat exchangers transfer as much as 75% of the heat from the stale air that is exhausted from the house to the controlled flow of air entering the house. The incoming fresh air reaches a temperature only a few degrees short of the household temperature. As an added benefit, a heat recovery ventilator also exhausts excessive moisture (from showers and cooking, for example) that can promote the growth of fungi and moulds.

Heat recovery ventilators work equally well in new energy-efficient homes and in conventional homes that have been retrofitted for energy efficiency. In both, the recovery of heat from exhausted household air means a savings of at least $100 a year. An HRV that is approved for installation in R-2000 houses is your best bet. Look for a unit that has an efficiency rating of at least 75% at 0°C. A heating and ventilation contractor will steer you in the right direction.

Checklist for Choosing a Contractor

The following tips are adapted from *Heat Pumps for Electrically Heated Homes* (an EnerMark publication). Ask whether the prospective contractor:

- is a qualified dealer for the make of equipment they sell;
- has units that meet CSA (or other appropriate) standards;
- has passed the manufacturer's installation and service training program;
- has certified refrigeration and electrical staff;
- guarantees the installation work;
- will supply reference of past installations;
- provides 24-hour emergency repair service;
- offers service contracts after the manufacturer's warranty expires;
- itemizes exactly what will be done (and when) in a written contract.

COMMON INDOOR POLLUTANTS	
Pollutant	**Source**
Formaldehyde gas	Urea formaldehyde foam insulation, plywood, particleboard, carpets, furniture
Household chemicals	cleaning products, paint and paint solvents
Carbon monoxide, carbon dioxide, nitrogen oxides	furnaces, ranges, dryers, fireplaces, woodstoves

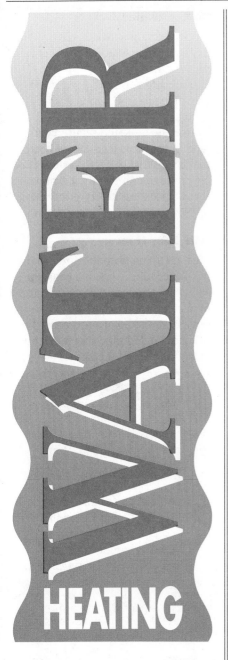

WATER HEATING

Domestic water heaters are the second-heaviest users of energy in the average Canadian home. Between 17% and 20% of the energy consumed by most families goes to ensuring that an abundant supply of hot water is always available at the turn of a tap.

Conservative estimates of the daily hot-water use of a typical family of four range from 125 L to 250 L. Chances are pretty good, however, that your family exceeds that limit just getting through its morning rush-hour routine. Running three five-minute showers, preparing breakfast, and turning on the dishwasher, for example, consumes 225 L of hot water – at an energy cost of 8.7 kWh of electricity.

HOW MUCH HOT WATER DOES YOUR FAMILY USE EVERY DAY?

Use	Litres
15-minute shower	40-160
Bath (half-full tub)	34-43
Whirlpool bath	400-1,200
Personal use (per day)	15
Hand dishwashing	7-16
Dishwasher	42-65
Laundry (1 hot-wash, cold-rinse load)	87

(Source: Ontario Hydro)

A sizeable portion of your energy-saving efforts should focus on reducing your family's hot-water demands. A quick shower uses a lot less water than a bath in a full tub, for example, and the warm-wash cycle gets clothes just as clean as hot water with less than a third of the energy. Running the hot-water tap is costly; for miscellaneous chores, like rinsing hands and dishes, reach for the cold-water tap and be sure to plug the sink first. When you need a basin full of warm water, add hot to cold rather than cold to hot.

an electric KETTLE

heats water far more efficiently than a pot or non-electric kettle. Fill it with only the amount of water you need, then switch it off as soon as it boils – or better, use a kettle that shuts off automatically when the water boils. Also descale the kettle regularly (rinsing well afterwards) to keep it operating efficiently.

ENERGY-EFFICIENT HOT-WATER TANKS

Replacing a hot-water tank is an expensive proposition. The first thing to sort out is matching the size of the tank to your family's needs. If each member of a family of four makes some energy-saving efforts and heat loss is cut to a minimum, a 180-L tank should be sufficient. The next step is to shop around for the most energy-efficient tank you can afford.

Performance standards for combustion units (gas and oil) are still being developed, so there's no guarantee that one type is more efficient than another. It's fair to say, however, that forced-draught and condensing-combustion hot-water heaters are the most efficient – and the most expensive.

Performance standards already exist for electric hot-water heaters sold in Canada. Each electric tank that meets the CSA C-191 performance standard carries a blue CSA (Canadian Standards Association) sticker that includes an efficiency rating. A good rule of thumb when purchasing an electric hot-water tank: a 180-L tank should have a rated standby loss of no more than 100 watts per hour; the rating for a 270-L tank should be no more than 115 watts per hour.

After reducing your family's hot-water demands as much as possible, turn your attention to increasing the efficiency of your hot-water tank. Here are a few cheap and simple tricks.

▬ Water Temperature

Check the thermostat on your hot-water tank. The setting should not exceed 54°C to 60°C. Unless you have a dishwasher requiring hot water at 60°C, you may be able to lower the temperature even further.

▬ Pipe Insulation

Long stretches of hot-water pipes, especially those that pass through unheated areas, are major sources of heat loss. A quick and easy solution is pipe insulation, which is readily available in larger hardware stores.

▬ Insulating Jackets

Hot-water tanks usually have 50 mm to 75 mm of insulation, but that won't prevent the tank from losing heat. You can add another 37 mm to 55 mm of insulation by purchasing a fibreglass jacket that is wrapped around the tank and secured with tape. If you have a gas water-heater, make sure that the blanket does not cover the opening at the bottom and that the jacket is taped tightly enough to prevent it from slipping down.

▬ Heat Traps

Heat escaping from the tank up the hot-water supply line is a constant source of heat loss. Ask your plumber to install a heat trap, which is a simple device consisting of a few short pieces of sharp-angled pipe that traps the heat at the top of the tank.

ABOUT HOME W·A·T·E·R TREATMENT SYSTEMS

About 100,000 home water treatment devices are sold in Canada each year to householders who are concerned about the quality of their tap water. Unfortunately these costly and totally unregulated devices can't deliver what consumers expect of them, and many actually create more problems than they solve.

The variety of sources of impurity in water is so broad that no single system can remove all the undesirables: microorganisms such as bacteria, viruses, and parasites; metals such as lead, iron, and mercury; organic pollutants such as nitrates from fertilizer run-off; toxins and carcinogens such as benzenes, pesticide residues, and trihalomethanes.

The most common element in treatment systems is a carbon filtration unit. Filters are not very effective against organic pollutants; more alarmingly, they can actually increase the bacteria count in the water passing through them. There is some evidence that they can also increase the level of trihalomethanes. These unwanted results can be moderated if the filter cartridges are replaced when they become ineffective, but there is generally no way the consumer can tell that this point has been reached.

Similar problems plague reverse osmosis devices. The best ones do a good job of removing microorganisms and inorganic salts, as long as their internal membranes remain intact. Most units have no way to determine whether the membranes or the seals around them have been punctured, and even tiny holes dramatically reduce effectiveness. In addition, reverse osmosis devices return to the user only 10% of the water that flows through them. The rest is sent down the drain – a tremendous rate of water waste.

There are other types of water treatment systems – ozonators, distillers, ceramic filters, even ultraviolet irradiators – but none has so far been invented in which the benefits clearly outweigh the risks and costs, in financial, health, or environmental terms. If you're willing to spend hundreds or even thousands of dollars for better water in your home, why not invest some of that money (and some of your time) in the battle against pollution. Pressuring industry and government to clean up the sources of water contamination and improve treatment of the public water supply will help to assure good water quality for your whole community.

The water used in Canadian homes accounts for 44% of municipal water demand. That's 1% more than the combined water demand of the industrial, commercial, and public sectors. The remaining 13% is lost to leakage.

Toilets and showers consume most of the water that municipalities pump into the average Canadian household. But the most notorious, not to mention most annoying, waster of water is the leaky tap.

* *

Leaky Taps

A tap leaking only one drop per second wastes more than 25 L of water a day, or 9,000 L a year. The problem is often a worn-out washer, which costs less than 10 cents to replace.

* *

Water-Saving Toilets

A standard toilet consumes about 20 L of water with each flush. You can reduce that amount by placing a plastic-wrapped brick or a plastic jug (weighted with a few small stones) in the tank to act as a dam. There are also water-saving toilets now on the market that operate perfectly well using 11 L to 13 L per flush.

* *

Water-Saving Showerheads and Faucets

A great deal of hot and cold water is wasted by conventional showerheads and faucets; for most families, showers are the single largest user of hot water. If you have a run-of-the-mill showerhead, it is probably spraying water at a rate of 15 L (and as high as 30 L) a minute. Consider replacing it with a low-flow showerhead with a flow rate that ranges between 7 L and 10 L per minute and is CSA-certified.

If you're content to do without water while you're soaping up your hair, for instance, attaching shut-off valves between the showerhead and downspout will save even more water and energy.

DON'T SING IN THE SHOWER

This table shows the litres of HOT water used in the average Canadian shower (by comparison, a half-full bathtub holds some 34-43 litres of hot water). A short shower under an energy-efficient showerhead can cut hot water dramatically.

	Standard Showerhead	Energy-Efficient Showerhead
5-minute shower	54 L	13 L
10-minute shower	110 L	27 L
15-minute shower	160 L	40 L

* * * * * * * * * * * * * * * * *

According to Ontario Hydro, one energy-efficient showerhead can save more than 28,000 L of water over the course of the year.

* * * * * * * * * * * * * * * * *

WATER METERING

Flat rates for water use encourage waste because they seem cheap: the rate is pitched unrealistically low, while the real costs are hidden in taxes. Studies have shown that installing water meters in homes, even without rate increases, permanently reduces water use 10% to 40%. An example: Edmonton meters all residential water, while Calgary is only partially metered. The result? Edmontonians use half as much water as Calgarians do.

If you care about throwing too much good water down the drain, consider lobbying for municipal water metering.

* * * * * * * * * * * * * * * * *

NOW AND INTO THE FUTURE
R·E·N·E·W·A·B·L·E E·N·E·R·G·Y

As conventional energy resources dwindle, the renewable energy technologies — solar, wind, mini-hydro, wood, and others — become increasingly important. Some environmentalists estimate that by 2025 Canada could be getting more than 80% of its energy from renewable sources.

Most homes in Canada are already solar-powered, at least partially — about 20% of a building's winter heating needs are supplied by the sunlight shining through south-facing windows. Known as passive solar heating, this is the simplest way of harnessing the sun's energy. Passive and R-2000 houses trap passive solar energy by focussing on these measures: selecting a site that has maximum solar exposure and summer shade; concentrating window space on the south side of the house and minimizing the number of north-facing windows; installing high-performance windows; insulating to the maximum level to prevent heat loss; designing the layout so that primary living areas receive as much direct sunlight as possible; planting evergreens on the north side of the house to protect against winter winds; and planting deciduous trees on the south side to provide cooling shade in the summer, while letting the sun shine through in the winter.

Anyone can follow the same principles, whether renovating an older home or building a new one. Before calling in an architect or drawing up your own plans, get in touch with your provincial energy ministry or hydro supplier and ask for information on passive solar design.

Selected sites across Canada — Prince Edward Island, Cape Breton, the southern Prairies — have winds high enough to justify large-scale windmill farms, where power can be generated

CONTINUED ON PAGE 77

It's the same story with faucets. Standard faucets flow at a rate of 11 L to 13 L per minute. Aerating faucets cut that rate in half, and the even more efficient units have cut flow rates to as low as 0.5 L per minute by using such water-saving technology as aeration, limiting maximum flow, and automatic shut-off. These features are unobtrusive, and it's likely you won't notice any difference in flow when you turn on the tap. Your plumbing wholesaler will be your best guide to water-saving faucets available in your area.

＊＊＊＊＊＊＊＊＊＊＊＊＊＊＊＊＊＊＊＊＊＊＊＊＊＊＊＊＊＊＊＊＊＊＊＊＊＊＊

MAJOR appliances

Few of us think of home appliances such as fridges, dishwashers, stoves, and as polluters, but in fact they are. Apart from the raw materials and energy consumed in their manufacture, and the pollution produced as a result, such machines use large amounts of electricity to run them, to heat water, and to pump water into them. The typical refrigerator, for instance, produces over 13 kg of acid pollution a year and as much as a fifth of a tonne over a 17-year working life.

The six major appliances – fridge, freezer, dishwasher, range, clothes washer, and dryer – account for 14% of the total energy used in the home. In a year, conventional (low-efficiency) models would consume a total of about 6,500 kWh of energy. But if all six were energy-efficient, you could cut that figure by 50%. In dollar terms,

that represents a saving of about $260 (assuming energy costs 8 cents per kWh, the national average). Over the lifetime of these appliances, savings could amount to more than $4,000.

L·I·F·E EXPECTANCY ·OF· APPLIANCES

dishwashers	13 years
clothes washers	14 years
refrigerators	17 years
ranges (electric)	18 years
dryers (electric)	18 years
freezers	21 years

＊＊＊＊＊＊＊＊＊＊＊＊＊＊＊＊＊＊＊＊＊＊＊＊＊＊＊＊＊＊＊＊＊＊＊＊＊＊＊

Energuide

Which is where Energuide comes in. In Canada, all models of the "big six" home appliances are tested for their energy use and rated in kWh per month. This rating must appear on an Energuide label attached to every new appliance (used appliances 10 years old or more should have the sticker too). The lower the figure, the more efficient the unit. This rating lets you compare models not only for energy consumption but also for actual price.

The energy cost represents a "second price tag" added to the purchase price of the appliance and should be part of your buying calculations:

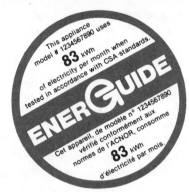

Second price tag = Energuide rating × 12 months × appliance life in years × local electricity costs (dollars/ kWh)

This figure can sometimes make a big difference in determining what is really your best buy. Refrigerator A, for instance, priced at $800 with a high efficiency rating of 84, would over its 17-year lifetime cost about $1,370 to run, for a total of $2,170. Refrigerator B, costing $700 but with a low efficiency rating of 145, would end up costing $3,070, or $900 more, while using nearly twice as much energy.

The Energuide Directory lists all "big six" appliances sold in Canada with their efficiency ratings. You can get a free copy (and back issues for used appliances) by writing to EMR Publications, 580 Booth Street, Ottawa, ON K1A 0E4.

✳✳✳✳✳✳✳✳✳✳✳✳✳✳✳✳✳✳

Refrigerators

After furnaces and hot-water heaters, refrigerators typically use the most energy in the home — up to 145 kWh per month. With the help of the Energuide label, you can buy more energy efficient units that cut that figure by almost half. The future will see models that offer even greater savings: *Wood's All Refrigerator Plus*, made in Guelph, costs the same as a conventional fridge but consumes only about 36 kWh per month (there is no freezer space, but consumers may select an optional ice compartment). And the California company

Sun Frost markets a more expensive, 16 cu. ft. model that uses a measly 20 kWh each month.

Besides choosing an energy-efficient model, there are some important tips to remember when buying and using a fridge:

▬ Buy the optimum size for your needs. A fridge that's too big obviously wastes energy, but one that's too small can use up extra energy too if it's always jam-packed. In general, 340 L (12 cu. ft.) is adequate for one or two people; 395 L to 400 L (14 to 17 cu. ft.) works for three to four people. Add 55 L (2 cu. ft.) for each additional person.

▬ Generally, the most energy-efficient style is a single-door, enclosed-freezer unit, while the least efficient is a two- or three-door unit with freezer chest beside the fridge compartment. There are exceptions to this rule — check the Energuide label.

▬ Use the energy-saver switch that lets you turn down or turn off the heating coils that warm the surface near the door opening (to prevent condensation).

▬ Always load or unload a fridge as fast as you can, so that a minimum of cold air escapes.

▬ Regular defrosting cuts running costs. Contrary to what you might think, frost buildup (more than 7 mm) can make your fridge less – not more – efficient.

▬ Adjust your fridge to the most energy-efficient temperature, using a thermometer if necessary: 3°C for fridge compartments, –18°C for freezers. Setting the temperature 4°C colder will increase the appliance's energy requirements by 10%.

▬ If possible, site the fridge well away from any heat source. At the very least, leave a good gap between them.

▬ Vacuum the cooling coils at the back of the fridge at least twice a year.

CONTINUED FROM PAGE 76

and fed into the local grid. Smaller, home-sized wind generators are ideal for meeting the electric power needs of remote sites where local hydro power is not readily available.

Solar electricity or photovolatics are also becoming more popular in isolated areas. Solar modules backed by battery storage systems can eliminate the cost and environmental destruction associated with bringing in power lines to remote properties. They can also be used to cost-effectively reduce or eliminate the running of diesel generators.

Heating an outdoor pool in Canada from May through September can use as much energy as heating a house through the winter. A solar pool heater can replace or reduce the use of a conventional pool heater, and pay for itself in four to six years. In new installations, the payback period can be as short as two years.

A typical solar domestic hot water system will supply between 40% and 50% of the yearly hot water requirements of a family of four. Depending on the area of the country and the cost of conventional fuel, the payback period for the solar system will be between four and seven years.

Throughout Canada there are thousands of sites where small hydroelectric — or mini-hydro — systems can be installed to generate electricity for communities of three to fifty houses. Even medium-sized streams with a water drop of less than 3 m can potentially produce all the electricity an average home needs. Many provincial power utilities will also buy any extra electricity that a mini-hydro site produces.

Almost 15% of all Canadian homes receive their primary heat from wood. Woodstoves can also be used as "zone heaters": heating the section of the house that's being occupied while

CONTINUED ON PAGE 78 ☞

CONTINUED FROM PAGE 77

reducing conventional fuel consumption overall. Inefficient woodstoves, however, can contribute to environmental and human health problems. Newer, high-efficiency stoves can reduce woodstove emissions by facilitating the secondary combustion of off-gases.

For more information on your renewable energy options, contact the following organizations:

❏ Biomass Energy Institute, 1329 Niakwa Rd., Winnipeg, MB R2J 3T4.
❏ Canadian Earth Energy Association, 228 Barlow Cres., Dunrobin, ON K0A 1T0.
❏ Canadian Photovoltaic Industry Association, Suite 3, 15 York St., Ottawa, ON K1N 5S7.
❏ Canadian Renewable Fuels Association, 190 Nicklin Rd., Guelph, ON N1H 7L5; or 201 – 4401 Albert St., Regina, SK S4S 6B6.
❏ Canadian Solar Industries Association, 67A Sparks St., Ottawa, ON K1P 5A5.
❏ Canadian Wind Energy Association, 44A Clarey Ave., Ottawa, ON K1S 2R7.
❏ Solar Energy Society of Canada Inc., 15 York St., Suite 3, Ottawa, ON K1N 5S7.

Buying a new fridge poses another environmental problem: what to do with the old one. When it's carted off to the municipal dumpsite, its CFC coolants are eventually released into the air. The solution may be new "vampire" technology that sucks out CFCs (they can then be cleaned and recycled or destroyed). Several municipalities plan to establish test programs to capture CFCs from so-called white goods such as freezers and refrigerators.

Energy experts do not recommend retiring the old fridge to the garage or basement to do service as a beer cooler — two small refrigerators require more energy to operate than one large one.

✳✳✳✳✳✳✳✳✳✳✳✳✳✳✳✳✳✳

Freezers

Most of the rules that apply to fridges also apply here. The most efficient models use about half as much energy as the least efficient units. And again size is important: don't buy a freezer that's too big for your needs. Generally, 130 L (4.5 cu. ft.) per person is about right.

Style is the next consideration, and here the choice is clear: chest freezers, with the door on top, are much more energy-efficient than upright models. Cold air leaks out around the door of an upright and rushes out every time you open it, but with the chest type, air tends to stay inside. Look for a unit that's well insulated: you can buy freezers with up to 7.5 cm insulation in the walls.

✳✳✳✳✳✳✳✳✳✳✳✳✳✳✳✳✳✳

Dishwashers

Key environmental question marks over dishwashers are the use of water and energy. According to a Science Council of Canada discussion paper on water, dishwashers account for an average of 1,200 L to 2,000 L of water per household per month. And during operation an average dishwasher uses about 100 kWh of energy per month, about 80% of it to heat the water (each load uses 42 L to 65 L of hot water). Using efficient appliances and techniques can net significant savings in both water and energy.

Detergent use is also important, though your dishwasher is likely to require little more detergent than you would use in washing by hand. In Canada, most automatic dish washing detergents are very high in phosphates, chemicals that can have an adverse effect on lakes and water life (see Chapter 3). Use a biodegradable, phosphate-free brand (see The Green Directory for a partial listing of available products) and experiment to find the smallest effective amount you can use under different circumstances.

When buying a dishwasher, compare the Energuide labels (ratings range from 83 to 121 kWh/m) and look for these features:

▬ A switch that turns off the heat-drying part of the cycle; air-drying works just as well and saves energy.

▬ A booster heater or "sani" setting that heats the incoming water to the recommended 60°C. You can then turn your water heater down to 55°C and save significantly in household water-heating costs.

▬ Extra-powerful spray action available on some newer units, which cuts down on water use.

Using your dishwasher wisely saves more energy. Try to wash only full loads and use the "econo" cycle when possible. Rinsing dirtier dishes by hand before loading can allow you to use the shorter cycle more often.

Ranges

Today's shopper faces a myriad of options in stoves – gas, propane, or electric, separate cooktops, self-cleaning and other special features, convection ovens, microwaves, and various configurations of any of these – that complicate the task of comparing specific models. Gas stoves are generally considered more energy-efficient than their electric counterparts (which use energy to heat up), but Energuide ratings are available for electric ranges only. Conventional electric stoves don't vary much in efficiency (59 to 73 kWh), but it pays to look for a low rating.

That said, these tips for choosing and using cooking appliances can help you save energy.

▬ The main feature to look for in a gas range is electronic ignition. Pilot lights waste energy by constantly burning a small amount of gas. If you have a gas stove with pilot lights, consider turning them off and using matches. Good insulation and a tightly closing oven door also save energy in gas stoves.

▬ Convection ovens are the most energy-efficient electric ovens available (although Energuide doesn't measure this feature). They incorporate fans that blow air around inside the unit, ensuring even heat and faster cooking at lower temperatures.

▬ The self-cleaning option can be an energy-saver: it requires intense heat, but since stoves with this feature tend to be better insulated, they use less energy in cooking. You save even more energy by cleaning only when necessary and then right after cooking, to take advantage of the heat.

▬ Smooth-surface elements use more energy than conventional ones.

▬ Built-in exhaust fans tend to be energy-burners in winter because they draw heated air from the room, causing the furnace to work harder.

▬ In cooking, use the right-sized pots (covered) and as little water as possible; don't overcook. The right small appliance for the job can save energy too – for example, cooking a stew in a slow-cooker rather than a saucepan consumes 80% less energy.

▬ Use your oven to best advantage: don't preheat unless necessary (such as for baking) and then only for about 10 minutes. Try to cook several dishes at a time. Turn off the oven a few minutes early and let the trapped heat finish the cooking.

* * * * * * * * * * * * * *

If you have a gas stove with pilot lights, consider turning them off and using matches.

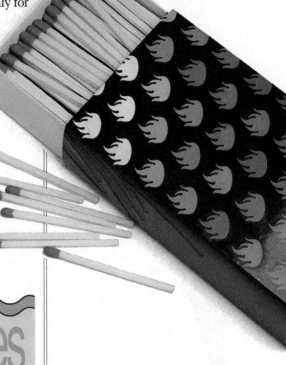

microwaves

Some 63% of Canadian kitchens now contain that supreme energy-saver the microwave oven. Not only does it save by accelerating cooking times, almost all the energy is used to cook the food, not heat the oven. For most cooking jobs, microwaves use less than half as much energy as do standard ovens. Most people who own a microwave oven, though, use a regular oven too. Your best bet could be a microwave with a convection feature, the most energy-efficient set-up of all. There is, however, an environmental problem associated with microwaves. The growing use of convenience foods that microwaves help encourage often means there are more packaging materials to throw away: boxes, disposable dishes, plastic wrap, and paper towels.

GARBAGE DISPOSALS

Garbage disposals waste water and electricity as well as putting more stress on the sewage treatment system. Compost heaps are much friendlier to the environment.

Clothes Washers

The same environmental factors applying to dishwashers - water and energy - apply here. But choosing carefully can make a dramatic difference to energy use and "second price tag" cost.

Energuide ratings vary from 54 to 136 kWh/m. Look for a low figure and these features:

▬ Front-loading machines (the laundromat type), which are common in Europe but can be hard to find here. They usually use much less hot water than top-loading models.

▬ Compact washers, which could save energy if your usual wash load is small; if not, however, you could end up using more energy by doubling up on loads.

▬ A cold wash and rinse cycle; using it as often as possible can reduce energy use significantly.

▬ Water-level control for small loads and shorter, gentler cycles for fine fabrics and less-soiled clothing.

* *

Dryers

By far the most energy-efficient method of drying clothes is, of course, to hang them up outdoors (in winter, try using indoor rack dryers). If you find that suggestion impractical, the next best option is to find the most energy-efficient dryer you can. Energuide ratings for dryers vary from 74 to 111 kWh/m, a smaller difference than for some other appliances but still worth checking out. Models being developed for sale in the mid-1990s that will operate using microwaves or heat pumps promise major energy savings. When you're shopping for a dryer, consider these factors too:

▬ A compact model saves energy if it's sufficient for your needs; if not, a full-size unit probably saves more.

▬ A sensor that automatically shuts off the machine when clothes are dry saves energy and clothes.

▬ A cool-down or "permapress" cycle that switches the air from hot to cool near the end of the drying cycle is an energy-saver.

▬ If you buy a gas dryer, choose one that has electronic ignition rather than a pilot light.

▬ A couple of usage tips: avoid partial loads and clean the filter between loads.

* *

Air Conditioners

Keeping the heat out of the house in summer is a problem many Canadians attack with air conditioning. Unfortunately, even the most efficient air conditioners available are major energy users, and a really super one can't be found. "Efficient" room air conditioners consume some 7,200 BTUs per hour, while central units use up 24,000. Air conditioners going full-blast in the heat of summer can, in fact, force electrical generating plants to pump so much carbon into the air that they actually heat up a city.

But many natural, non-polluting methods exist for cooling the home that are cheaper and quieter than air conditioners and often just as effective. Before capitulating to machinery and adding CFCs to the atmosphere, consider trying some of these "greener" keep-cool alternatives.

■■■ Plant trees. Not only do they shade the house in summer, they also provide insulation against wind and snow in winter. Climbing vines, especially on east and west walls, can help shade windows.

■■■ If you've installed heat pumps for heating, you've also got free air conditioning (see pages 71-72).

■■■ Consider installing low-emissivity (low-E) windows (see pages 68-69). Their almost invisible metal coating reflects away outside heat and again has the double advantage of helping keep heat in during winter.

■■■ Pull down the shades. Awnings (look for cotton-fibreglass fabric) on south-facing windows help keep out the high midday rays, and vertical shading is effective for west- and east-facing windows, which get the morning and evening sun at low angles.

■■■ Sunscreen blinds, made of fibreglass and polyvinyl, reduce the sun's glare and heat by 70% to 85%, according to manufacturers.

■■■ Install low-speed ceiling fans. A fan that moves air at only about 3 km/h can make a big difference in comfort and is actually more efficient than a higher-speed fan, whose additional cooling isn't sufficient to justify the increase in noise and energy use.

■■■ Then there are the common-sense measures: keep the doors and windows closed until evening; shade east windows in the morning, west windows in late afternoon; as often as possible substitute your microwave, barbecue, and small electric cooking appliances for stovetop and oven.

> Air conditioners going full-blast in the heat of summer can force electrical generating plants to pump so much carbon into the air that they actually heat up a city.

* *

LIGHTING

In most homes, lighting is not a major energy problem, accounting for 2% of total energy use (about 1,000 kW/h per year). But new technology can in some cases reduce that figure by 70% to 80% while providing other advantages, so it's well worth considering.

* *

Compact Fluorescents

The big innovation in home and office lighting is compact fluorescent fixtures. Gone are the days when fluorescent meant long bulky tubes emitting poor-quality, flickering blue light. New compact fluorescent bulbs are not only far more efficient than regular incandescent ones but provide pleasant, comparable light quality for most purposes, and last 10 to 13 times as long. Some can even be screwed directly into standard fixtures, though most require an adaptor that fits into the socket.

They are also more expensive — $10 to $12 if you have an adaptor, $19 to $25 including the adaptor and depending on the wattage — compared with less than a dollar for standard incandescent bulbs. However, this difference is usually cancelled out by huge savings on the second price tag, the energy cost.

For More Information on Home Energy Use

The federal government has prepared factsheets and booklets on various aspects of home energy management. In addition to its Energuide Directory, the Department of Energy, Mines and Resources (Home Energy Information, 580 Booth St., Ottawa, ON K1A 0E4) publishes other free publications, including:

Heating & Cooling with a Heat Pump
Keeping the Heat In

The Ontario Ministry of Energy (Consumer Publications, 56 Wellesley St. W., 9th Floor, Toronto, ON M7A 2B7) has published:

Where and How to:
❑ *Caulk & Weatherstrip*
❑ *Install Air-Vapour Retarders*
❑ *Insulate Basements*
❑ *Provide Fresh Air & Control Humidity in a Tighter House*
❑ *Insulate Cathedral Ceilings and Flat Roofs*
❑ *Improve Fireplace Efficiency*

Consumer's Guide to Buying Energy-Efficient Appliances and Lighting
Consumer's Guide to Buying Energy-Efficient Windows and Doors
Consumer's Guide to Buying an Energy- Efficient Resale Home

Your municipal or provincial utility company is often a good source of energy conservation advice. Ontario Hydro (700 University Ave., Toronto, ON M5G 1X6) will supply copies of:

Build Energy Efficiency into Your Renovations
Electric Heating — Operating Tips for Efficiency and Comfort
Wise Ideas for Efficient Summer Cooling
Heat Pumps for Electrically Heated Homes
Residential Water Heaters — Wise Use, Safety & Maintenance Tips.

The key to energy efficiency in lighting is how much light is actually produced (lumens) in relation to the amount of power consumed (watts). Incandescent and fluorescent bulbs vary enormously in their lumens-per-watt ratios. Compact fluorescents, for example, produce four to five times as many lumens-per-watt as incandescents do. In other words, an 18-watt fluorescent can do exactly the same job as a 75-watt incandescent, while lasting 10 times longer.

Compact fluorescents are available in a variety of futuristic styles – circular, U-shaped, "quad cluster" – as well as the familiar globe, some requiring an adaptor to plug into standard sockets. Before buying, be sure the lamp will fit the intended fixture; some may be too large.

Fluorescent technology is improving rapidly. Look for compact fluorescent bulbs that are unitized, incorporating a disposable electronic ballast built right into the base of the bulb. These newer units eliminate the annoying pulsing or flickering problem of earlier compacts with magnetic ballasts.

Finding compact fluorescents, however, can be a problem since they're not yet widely available. Check specialty lighting stores and electrical equipment outlets; if they don't have what you want, ask them to order it.

* *

Energy-Efficient Incandescents

Another option is a variety of new energy-efficient incandescent bulbs. For instance, tungsten halogen bulbs can cut power consumption by 50% over conventional ones and last up to two and a half times longer. In these bulbs, the tungsten filament is contained within a small quartz capsule filled with halogen gas, similar to an automobile headlamp. This gas enables the tungsten, which evaporates from the filament, to be redeposited back onto the filament instead of onto the inside surface of the bulb (the cause of darkening in regular incandescents). Consequently, the bulbs last longer and have a greater lumen output.

Parabolic-aluminized reflector (PAR) lamps, a tungsten-halogen type, can replace conventional pot or spot lamps in many decorative lighting applications in the home and will pay for themselves quickly. They last up to 6,000 hours and are available in strengths from 75 to 1,500 watts.

When you're buying incandescents, watch the terminology: "long-life" bulbs are in fact less efficient than standard incandescents, and "energy-savers" save energy by using lower wattage but aren't necessarily more efficient.

Lightbulbs aside, the way you use lighting can also make a big difference in energy savings:

▬ Use only as much light as you need for the purpose. A 25-watt incandescent bulb may work well for a porch light; save the bright lights for reading lamps and other special tasks.

▬ Consider installing energy-saving devices like timers, which automatically turn off lights at preset times, and photocells, which turn off night lights during the day.

▬ Install dimmer switches. Dimming is possible with fluorescents as well as incandescents, but ask for the newer bulbs with electronic ballasts. Unlike earlier types, they provide dimming without loss of efficiency. You can get them in several configurations: fixed dimming (preset at outputs of 85%, 70%, 50%, and 35%), manual dimming (with an output range of 100% to 35%), and automatic dimming (using programmed timers or photocell controls).

Batteries–Energy to Go

Battery manufacturers have done a lot in recent years to reduce the use of mercury and other heavy metals in their products; almost all alkaline manganese batteries now contain less than 0.025% mercury, while mercury has been virtually eliminated from most zinc carbon batteries. (Some batteries may still contain small amounts of mercury; read the package or label or contact the manufacturer for mercury content of a particular brand.)

Despite these improvements, disposable batteries are still a throw-away convenience product and, for all practical purposes, are non-recyclable. (It costs more to extract the contaminated raw materials from the used batteries than they are worth.) If Green Consumers need a battery, they try to pick the right one for the job, use it wisely, and dispose of it responsibly.

Battery	Use	Comments
Alkaline manganese	the most common small batteries in Canada, used in heavy power drains such as portable stereos, cameras, and camera flash attachments	disposable/non-recyclable
Zinc carbon	used in less demanding applications, such as radios, bicycle lamps, flashlights, shavers, clocks, calculators, and TV remote-control units	disposable/non-recyclable
Button cells	tiny batteries used in hearing aids, watches, and some cameras	disposable/non-recyclable; buy zinc air button cells rather than mercuric oxide or silver oxide cells
Lithium	last 2 to 3 times longer while costing twice as much as alkaline batteries	disposable/non-recyclable
Nickel-cadmium rechargeable	can replace disposables in most applications but need to be recharged more often when used in portable stereos and other high power drains	more expensive initially but can be recharged up to 500 times; must be used carefully or their life is shortened; contain cadmium

So What Do We Do about Batteries?

The answer is difficult. The best advice is probably:

▬ Use plug-in power rather than batteries when you can. Manufacturing batteries can take 50 times more energy than they produce.

▬ Second choice: use rechargeables, especially for equipment that's a heavy user of batteries.

▬ If you must use disposable batteries, pick the type most suitable to the appliance. And select mercury-free brands.

▬ Whichever batteries you use, don't mix new batteries with old. The new batteries try to recharge the old ones and their life is cut significantly.

▬ Batteries that no longer have enough power to operate a photoflash unit or other heavy energy drainer may still have enough left to drive a radio or portable cassette player.

TRY NOT TO BUY ENDLESS PACKS OF DISPOSABLE WORK GLOVES, whether for painting and decorating or for use in the kitchen or garden. Buy a sturdy, long-lasting pair, even if they're more expensive initially.

Each rechargeable battery saves between 150 and 250 regular batteries from being thrown away. However, rechargeable batteries contain up to 30% cadmium (by weight), a heavy metal that causes kidney and liver damage, emphysema, anaemia, bone disease, and other ailments. Cadmium is also a possible human carcinogen. Battery manufacturers are still some years away from developing a non-cadmium-based rechargeable. Until that time, Green Consumers should press government and industry to set up a collection system for old rechargeables — and keep them out of our incinerators and landfills.

****

INDOOR AIR POLLUTION

T he average Canadian spends approximately 90% of the day indoors, breathing the same stale, polluted air in and out, in and out. And getting sick. That's because the typical home is full of dangerous chemicals — they're in paints and glues and solvents, putty and floor sealers, upholstery fabric and foam, carpets and wallpaper. They escape from improperly maintained or vented heating and cooking systems. They leak in from the outside air and soil. And they drift off the burning ends of cigarettes and pipes.

The problem is made worse by the tendency of pollution to become concentrated in our well-insulated, airtight modern houses and apartments. The air indoors often contains two to five times the level of toxic chemicals found in the air outside. The toxic contamination can even exceed the workplace safety limits that apply to offices and factories.

The short-term health effects — tiredness, sore throats, burning eyes, and so on — have recently been recognized as the exhausting and often debilitating condition known as "sick building syndrome." People with chronic health problems such as asthma, heart disease, cystic fibrosis, and AIDS may be especially susceptible. The long-term implications of exposure to toxins such as formaldehyde, asbestos, second-hand cigarette smoke, and radon can be equally serious and, in some cases, fatal.

A number of factors affect indoor air quality:

- the materials used to construct the home
- the materials used to furnish and decorate the home
- the lifestyle of the homeowners (particularly whether they smoke)
- the products used to cook and clean
- the heating and ventilation system
- outdoor pollution (it can be sucked inside by ventilation equipment)

Smoking is probably the single greatest contributor to indoor air quality problems; an estimated 50,000 tonnes of tobacco are burned indoors by Canadians each year, releasing more than 4,700 chemical compounds, including 43 known cancer-causing agents. Fortunately, cigarette smoke is also one of the easiest air pollutants to control — simply ban smoking in your home.

Other serious pollutants include: radon gas (a natural pollutant that seeps into and builds up in modern airtight homes); carbon monoxide (which kills dozens of Canadians each year) and the other products of incomplete combustion; formaldehyde; asbestos and other fibres; various bacteria and viruses (that, along with fungi and moulds, thrive in damp environments); pesticides and other toxic organic compounds.

FOR MORE INFORMATION

The federal government has published a series of booklets and reports for consumers concerned about indoor air pollution. Contact the Canadian Mortgage and Housing Corporation (Publications, 682 Montreal Rd., Ottawa, ON K1A 0P7) for copies of:

How to Improve the Quality of Air in Your Home ($2)
Ventilation — health and safety issues ($1)
Guide to Radon Control ($2)
Radon Control in New Housing ($2)
Moisture and Air — Problems and remedies (free)
Research Reports and Papers on Indoor Air Quality (a free checklist)

Health and Welfare Canada (Communications Branch, Jeanne Mance Building, 19th Floor, Ottawa, ON K1A 0K9) has published *Exposure Guidelines for Residential Indoor Air Quality*. And the Lung Association (573 King St. E., Toronto, ON M5A 1M5) has published a series of pamphlets on indoor air quality and common air pollutants.

Common Indoor Pollutants

Pollutant	Source	Health Effects
Tobacco smoke, containing more than 4,700 chemical compounds, including carbon monoxide, tar, nicotine, PAHs, nitrosamines, arsenic compounds, formaldehyde	cigarettes, pipes, cigars	eye, nose and throat irritation, headaches, bronchitis, pneumonia, respiratory and ear infections in children, lung cancer, heart disease; second-hand smoke may increase the risk of lung cancer by 30% in non-smoking partners, and wheezing and coughing by 20% to 80% in the children of smoking parents
Carbon monoxide (CO)	a combustion by-product, CO can be produced by furnaces, ranges, dryers, fireplaces, woodstoves, automobile exhaust, tobacco smoke	fatigue, chest pains, impaired vision, dizziness, confusion, nausea, flu-like symptoms; fatal at very high levels
Nitrogen oxides (NOx)	a combustion by-product, generated by kerosene heaters, unvented gas stoves and heaters, tobacco smoke	eye, nose, and throat irritation, impaired lung function and respiratory infection in young children
Radon	a "natural" pollutant, radon migrates from earth and rock beneath the home; small amounts may be released from well water and some building materials	long-term exposure to radon gas is suspected to be the second most significant cause of lung cancer (after tobacco smoke)
Formaldehyde gas	urea formaldehyde foam insulation (UFFI), pressed wood products (plywood, particleboard, fiberboard), furniture, textiles, tobacco smoke, some glues	irritation of eyes and throat, coughing, fatigue, skin rash, severe allergic reaction, cancer
Volatile organic gases	paints, paint strippers, and other solvents, wood preservatives, aerosol sprays, cleansers and disinfectants, air fresheners, fuels and automotive products, etc.	eye, nose, and throat irritation, headaches, loss of co-ordination, nausea, damage to liver, kidneys, and central nervous system; some organics are suspected carcinogens
Lead	lead-based paints (common before 1970), lead solders in water pipes, lead batteries, ammunition, automobile exhaust, some glasses, insecticides, ceramic glazes, and rubber compounds	impaired mental and physical development in young children; damage to red blood cells, kidneys, and nervous system; miscarriages, sterility, and sexual dysfunction

The Health Effects of Sick Building Syndrome

- irritated or burning eyes
- fatigue
- lingering colds and stuffy nose
- nose bleed
- sore throat and hoarseness
- wheezing and coughing attacks
- constant headaches
- itchy, dry skin or rashes
- nausea and dizziness
- muscle aches and pains

The mental symptoms are most difficult to isolate, but may include bouts of lethargy or hyperactivity, irritability, nervousness, mental fatigue, and difficulty concentrating.

None of these symptoms is unique to sick building syndrome. However, the problems may be related to indoor air pollution if they are shared by other family members; other causes have been eliminated; symptoms clear up when you spend time outdoors; or you feel better when you leave the windows open.

Asbestos	insulation, fireproofing, and acoustical materials; various construction materials	lung diseases, chest and abdominal cancers
Pest control products	lawn, houseplant, and garden care products, insecticides and herbicides, flea powders for pets, etc.	irritants, central nervous system damage (headache, dizziness, numbness and weakness, loss of co-ordination, mood swings), cancer; high levels can be fatal
Ozone	frayed electrical connections, improperly maintained electrostatic air filters, outdoor sources	irritation of nose and throat, vision problems, heart disease, damage to red blood cells
House dust/ suspended particulate	mineral, wool, and glass fibres, silica dust, pet hairs, human skin, house mites and insects, pollens and spores, tobacco smoke, outdoor sources	irritation of eyes, nose, and throat, allergic responses, respiratory disease
Infectious agents (viruses, fungi, and bacteria)	human carriers, damp rugs and wall, rotting woodwork, air conditioners, humidifiers, toilets	the increased and localized incidence of dangerous illnesses, including Legionnaire's disease, measles, Q-fever, and influenza, spawned by unsanitary conditions, gave rise to the term "sick building syndrome"

Test It Yourself

You can hire an environmental testing firm to assess the pollutant hazards in your home (contact your local health department for labs in your area). There are also do-it-yourself test kits on the market for measuring specific contaminants.

Lead: Lead-Check and Lead-Check II (Abotex Enterprises, 105 Christina St. S., Box 416, Sarnia, ON N7T 7J2); Test for Lead and Other Metals in Pottery (Frandon Enterprises Inc., P.O. Box 300321, Seattle, WA 98103);

Carbon Monoxide: Quantum Eye (Quantum Group Inc., 11211 Sorrento Valley Road, SuD, San Diego, CA 92121); Backdraft Detectors, (Alpha Detectors, 515 Princess Louise, Orleans, ON K4A 1Y2);

Radon: The Green Directory at the end of this book lists radon detection firms that can supply you with a simple test kit.

****

Chemical Hypersensitivity

The discussion and product choices in this book focus mainly on environmental impact, with health issues sometimes briefly mentioned. Some of our choices, however, would be inappropriate for people with allergies or special chemical sensitivities. An estimated 30,000 Canadians suffer from chemical hypersensitivity, which has also been called total allergy syndrome, environmental illness, or simply 20th-century disease. The problem usually begins with an adverse reaction to a single chemical, but may progress to a complete breakdown of the immune system. Even natural substances such as cotton and wood, for example, which are often suggested here as preferable from an environmental viewpoint, can be problems for certain people.

If you have or suspect you have environmental allergies, contact one of the following national organizations (there may be branches in your province) for more information:

❏ AGES, Advocacy Group for the Environmentally Hypersensitive (1887 Chaine Court, Orleans, ON K1C 2W6)

❏ Allergy and Environmental Health Association (10 George St. N., Cambridge, ON N1S 2M7)

❏ Allergy Foundation of Canada (P.O. Box 1904, Saskatoon, SK S7K 2S5)

❏ Allergy Information Association (65 Tromley Dr., Islington, ON M9B 5Y7)

❏ Canadian Society for Environmental Medicine (RR 6, 6901 2nd Line West, Mississauga, ON L5M 2B5)

❏ Parents of the Environmentally Sensitive (151 Sutherland Dr., Toronto, ON M4G 1H8)

✱✱✱✱✱✱✱✱✱✱✱✱✱✱✱✱✱✱✱✱✱✱✱✱✱✱✱✱✱✱✱✱✱✱✱✱

How to Control Indoor Air Pollution

Methods of controlling and improving indoor air quality include reducing or removing the source of the problem, cleaning and filtering air, humidifying and dehumidifying, installing local exhaust fans, and guaranteeing good general ventilation.

Whenever possible, eliminate the source of the toxic contamination. The environmental and health problems associated with furnishing and decorating your home usually can be summed up in two phrases: "bad" woods and hazardous chemicals. The cure is equally straightforward, though not always easy to apply: use "good" woods (see page 91), and choose natural over synthetics.

▰ If possible, don't store toxic products such as paints, pesticides, gasoline, and other fuels indoors.

▰ Carefully seal all chemical products.

▰ Clean up all chemical spills immediately.

▰ Confine hobbies that require concentrated chemicals to a well-ventilated area.

▰ Ban smoking in your house. Or confine it to one separate, well-ventilated area.

▰ Fuel-burning equipment, such as furnaces, hot water heaters, gas-fuelled clothes dryers, or woodstoves are designed to be vented outdoors.

▰ Have your heating equipment, fireplaces, and chimneys regularly inspected and serviced.

▰ If you have replaced your electric stove with a gas one for reasons of energy efficiency, you must install fan ventilation (vented directly outdoors) over your gas range. Contact a heating/air conditioning contractor for advice.

▰ Clean air conditioners, air ducts, humidifiers, dehumidifiers, and heat exchangers regularly. Keep humidifiers and air conditioners free of mould and mildew. (Cleaning with mild borax solution can be effective.)

▰ If you use firewood, dry it outside and bring in only the quantity you need.

▰ High heat and humidity can stimulate the release of volatile chemicals from household furnishings and building supplies.

▰ Clean furnace filters regularly (once a month or so) and replace when necessary.

▰ Air filters can be used to remove certain contaminants: electrostatic medium-efficiency pleated filters can remove microscopic particles from the air (they do not remove volatile chemicals or gases); activated carbon filters or chemical absorbent filters can remove some gases; some houseplants can remove organic chemicals.

▰ Do not rely on small "desk-top" filters and devices to keep the air clean — they do not effectively remove toxic gases and small particles.

CFCs ON THE WAY OUT

Canada's program to ban ozone-destroying chlorofluorocarbons (CFCs) is speeding up. Canada has committed itself to a complete phase-out of CFCs by 1997, three years ahead of the international targets agreed to in the Montreal Protocol on Substances that Deplete the Ozone Layer. Federal regulations that took effect January 1, 1991, ban the manufacture, import, or sale of:

• aerosol cans containing CFCs.
• small cans of CFC refrigerant used by do-it-yourselfers to recharge air conditioners, and
• plastic foam food or beverage containers made with any CFC foaming agent.

Future regulations will ban the use of CFCs in rigid foam board insulation by 1994. Other uses — including CFCs in new refrigeration units, soft foams and other rigid foam applications, and home and automobile air conditioners — will be the subject of ongoing discussions.

Other chemicals being phased out under the Montreal Protocol are halons (used in fire extinguishers), carbon tetrachloride (used in the production of CFCs, pesticides, and solvents), and methyl chloroform (used as an industrial solvent to clean metal and electronics and as a carrier solvent for adhesives). Several provinces and even some municipalities have placed their own controls on the use of ozone-destroying compounds. For more information, contact the Commercial Chemicals Branch, Environment Canada, Ottawa, ON K1A OH3.

A dedicated insulated air duct can be used to supply combustion air directly to fuel-burning heating equipment and appliances. Unless adequate air is supplied to such equipment, negative pressure conditions can develop in the home, the furnace will burn inefficiently, carbon monoxide and other dangerous combustion gases will be drawn back down the chimney, and radon gas will be sucked in through cracks in the basement.

Inspect heat exchangers regularly and keep them in good repair; a cracked or perforated exchanger can allow combustion gases to enter the home.

Vacuum dust from refrigerator coils on the back of the appliance.

Control radon levels by sealing all cracks and holes in the below-ground level of your home. Other entry points include water and sewer pipes, floor drains, construction joints (where floor and walls meet), and up the jack posts that support the floor beams.

Never leave a car or lawn mower running in an attached shed or garage.

Never light or cook on a charcoal or gas-fired barbecue inside your home, garage, or trailer.

* *

A Truly Green Air Filter

The leaves of houseplants filter and clean the air, and give back fresh oxygen. Studies conducted by NASA suggest that houseplants can reduce the levels of at least three common indoor air pollutants — formaldehyde, benzene, and trichloroethylene. Only a few plants and pollutants have been tested so far, but look at it this way: the beauty of a little more greenery can't hurt. An if it happens to make breathing a little easier too, that's a bonus.

NASA researchers estimate that one 30 cm plant per 9 square metres of floor space should be sufficient for cleaning at least some of the contaminants from household air. The following plants are found to be effective in removing the three chemicals used in the tests:

- heart-leaf philodendron
- lacy tree philodendron
- spider plant
- snake plant or mother-in-law's tongue
- pot mum
- aloe plant
- bamboo or reed palm
- Janet Craig dracaena
- Warneckeii dracaena
- Gerbera daisy
- dragon tree
- golden pothos or devil's ivy
- English ivy
- banana tree
- peace lily or spathe flower

Who to Contact for More Information on Renovation Hazards

The local provincial ministry of labour office, occupational health and safety branch, or Workers' Compensation Board will be able to answer some of your questions on renovation hazards. Additional information is available from:

Canadian Centre for Occupational Health and Safety, 250 Main St. E., Hamilton, ON L8N 1H6; or call toll-free 1-800-263-8267

Canadian Safety Council, 1765 St. Laurent Blvd., Ottawa, ON K1G 3V4

Construction Safety Association of Ontario, 74 Victoria St., Toronto, ON M5C 2A5

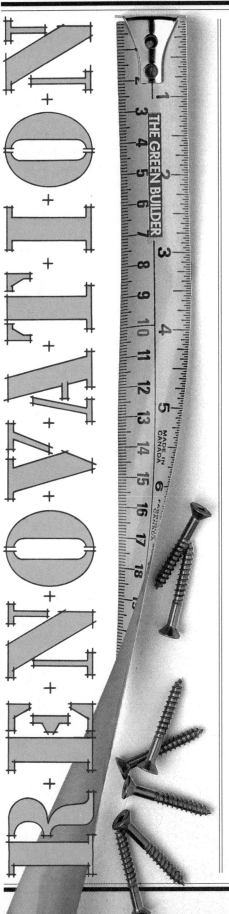

RENOVATION

Despite the hard work and hassles, few projects are ultimately more satisfying than renovating a home. But almost everything you tear down and everything you put up in its place is potentially harmful to the environment or to you. Modern homes are largely composed of synthetic materials, many of which are based on non-renewable resources or use these resources in their manufacture. Many require tremendous amounts of energy to produce. And many involve the use of toxic chemicals that can create pollution problems at every stage, from manufacture to disposal.

These chemicals don't like to stay put. Everyone is familiar with that "new car smell." It is caused by chemicals evaporating or "outgassing" from the plastic seat covers, rubber floor mats, and all the other synthetic components in the vehicle. The same thing takes place at home after you've painted, or laid a new rug. Although we usually can't smell them, dozens of dangerous pollutants continue to leak out of modern building materials for at least 30 years after a house is constructed.

As an overall guideline for home construction, renovation, or repair: choose the natural rather than the synthetic, the product based on renewable rather than non-renewable resources.

* * * * * * * * * * * * * * * * * *

Protect Yourself Against These Chemical Hazards

Hazard: **Asbestos dust**
Source: Vinyl floor tiles and acoustic tiles, cement pipe, roofing felts and shingles, asbestos siding, insulation around heating ducts
Prevention: Arrange for removal by licensed contractor (equipped with appropriate protective clothing and breathing equipment).

Hazard: **Coal tar products**
Source: Roofing, waterproofing
Prevention: Wear proper breathing equipment, gloves, and barrier creams.

Hazard: **Wood preservatives**
Source: Cutting prepared wood, applying preservatives to cut ends
Prevention: Wear gloves, an apron, safety glasses, and a mask/respirator to protect against dust and fumes; do not burn scrap.

Hazard: **Solvents**
Source: Adhesives, gluing and cutting PVC pipe, caulking and sealing compounds, paints and paint strippers
Prevention: Keep containers tightly sealed; wear gloves, protective clothing, and appropriate respirator (for protection against organic vapours); ensure good ventilation; remove all sources of sparking or ignition.

Hazard: **Epoxy resins**
Source: Laying tiles, certain paints, "super" glues
Prevention: Wear gloves and hand barrier creams; ensure good ventilation; keep containers sealed.

Hazard: **Fibre particles**
Source: Glass fibre, mineral wool and cellulose fibre insulation
Prevention: Wear dust mask, gloves, and protective clothing; tape sleeves and trouser legs; launder work clothes separately.

A potential hazard for renovators of homes built between 1900 and the early 1970s is asbestos, a then-common insulating, fireproofing, and strengthening agent used in roofing shingles, siding, pipe and boiler insulation, sheet and tile flooring, ceiling tiles (for soundproofing) – even as sprayed-on decoration in ceilings or walls. Asbestos, now infamous for causing lung cancer and asbestosis (there is no known safe exposure limit), can escape in microscopic fibres that hang in the air if the source is disturbed in any way. If you intend to remodel, it must be removed.

If you suspect you may have asbestos anywhere in your home, contact your provincial health department for advice on identifying it and arranging for a trained, licensed asbestos contractor to handle it safely for you.

Lumber and Wood

F ew home renovators will choose a tropical hardwood for framing or a new deck. But if you're tempted by teak cabinets or doors, first read "Good Woods" (page 91). For wood used in basic construction, the main environmental issue is not tropical deforestation but hazardous chemicals.

Wood windows and door frames are often treated with poisonous insecticides, mildewcides, fungicides, or other chemicals that can outgas (give off fumes) into your home for months or even years. When disposed of as wastes, either during manufacture or after use by the consumer, these chemicals may contaminate water or soil.

An especially noxious chemical, formaldehyde, is an ingredient of the binding agent in indoor plywood, chipboard, and particleboard. Formaldehyde can cause cancer and affect the central nervous system. Dimensional lumber is safer, although more labour is required to use it.

However, manufactured lumber does have the advantage of making use of wood that would otherwise be wasted. If you choose manufactured wood, there are a few things to remember. Use exterior-grade plywood, since its binding agent is more stable and will cause less outgassing. You can prevent outgassing entirely by sealing the lumber. Don't burn the scraps.

Whatever framing lumber you choose, you can economize on the amount needed by using drywall clips, trusses, and other framing innovations.

* * * * * * * * * * * * * * * * * * *

Decks

Pressure-treated lumber is resistant to decay caused by moisture and insects, but the preservatives are highly toxic. If you do use preserved wood for a deck or fence, take precautions. Don't use it near a vegetable garden; wear protective clothing, a face mask, and gloves if you're treating cut ends with preservative; attach it with brackets to a concrete footing rather than sinking the wood directly into the ground where chemicals can leach into the soil; and don't burn the scrap.

Cedar, which is naturally decay-resistant, is an excellent alternative. Other softwoods work well outdoors, too, if you protect them with appropriate paint or stain. Protect outdoor wood from direct contact with the soil (attach it to concrete) to avoid wood rot.

* * * * * * * * * * * * * * * * * * *

Roofing

Slate, tile, metal, fibreglass, and asphalt roofing materials do not outgas chemicals, but they are all based on non-renewable resources. Of these, asphalt shingles, which are petroleum-based, have the shortest service life. Cedar shingles, though relatively expensive, are one roofing material based on renewable resources.

* * * * * * * * * * * * * * * * * * *

Siding

As is so often the case in trying to make environmentally sound buying decisions, the consumer faces tradeoffs in choosing siding. Siding falls into four types: masonry, metal, plastics, and wood. The first three are all based on non-renewable resources.

Metal and vinyl sidings, though they have the advantage of being maintenance-free, are extremely difficult to repair if damaged. Aluminum is very energy-intensive in its production.

That leaves wood – probably cedar, which weathers well in most climates. But check local fire regulations: if non-combustible materials are required, choose stucco, brick, or stone.

Flooring

Avoid such flooring materials as asphalt or vinyl tiles, which are based on petrochemicals, and choose those that are based on renewable resources and that require less energy to produce. Wood is your best choice. Other acceptable natural materials are ceramic tile, terrazzo, stone, brick, and quarry tile.

* * * * * * * * * * * * * * * * * *

Cabinets, Counters, and Interior Trim

For cabinetry, avoid particleboard and interior-grade plywood, which are probable sources of formaldehyde. If you have inherited some (check the sides, tops, bottoms, and shelves of "wood" cupboards and the base of countertops), seal exposed surfaces with special environmental sealers. Alternative cabinet materials are solid wood or metal.

Tropical hardwoods such as teak and mahogany are often used in cabinetry, doors, and interior trim. Don't use them unless you can determine that they've come from managed forests (see "Good Woods," below). Also avoid plastic for baseboards and other trim; it's derived from petrochemicals. Use native wood for trim.

* * * * * * * * * * * * * * * * * *

Disposing of Wastes

Any kind of home repair or renovation creates a lot of waste. Getting rid of it in a way that causes the least environmental harm can be difficult. Don't throw unwanted or waste paints, solvents, varnishes, wood preservatives, or glues down the drain or into the garbage, where they can eventually leach into soil or water. Dispose of them as hazardous wastes at household hazardous waste depots or through special municipal collection programs (see Chapter 7).

The same goes for scrap manufactured and treated wood. Never burn it, which sends chemical fumes flying into the air.

Good Woods

One of the key issues for Green Consumers is tropical deforestation. North America, Europe, and Japan are the major consumers of tropical hardwoods, most of which come from rainforests in the Philippines, Malaysia, Indonesia, South America, and West Africa. Every year millions of tonnes are exported to timber-hungry nations to make furniture, doors, window frames, construction materials, boats – even coffins.

Extracting the valuable timber species is a destructive process that also fells some nine times as many unwanted species, which are simply left on the ground; that compacts delicate forest soil and damages the roots of standing trees. Tropical forests are rapidly disappearing and with them the wildlife they sheltered, the soil they protected, the air moisture they produced. By the year 2000 it is expected that half the world's tropical forest will have been razed to supply the timber trade and make room for agriculture.

So what choices are there for the environmentally conscious Canadian consumer? Encourage the tropical timber industry to switch to sustainably grown tropical hardwoods by purchasing those woods, and use softwoods or temperate hardwoods whenever possible. Unfortunately, it's often difficult and sometimes impossible to determine the precise source of tropical hardwoods: for instance, some teak from Java, Thailand, and Burma, and some greenheart from Guyana is grown on sustainably managed plantations, but as yet you can't find out which. Malaysian rubberwood is a sure choice.

Look for temperate hardwoods like maple, cherry, oak, alder, apple, aspen, beech, birch, elm, hickory, and black walnut. Some softwood alternatives to hardwood are pine, spruce, hemlock, and Douglas fir. Where durability is a priority, however, the timber industry tends to treat softwoods with pentachlorophenol (PCP), Lindane, tributyl tin oxide (TBTO), and Dieldrin; ask for untreated.

use "GOOD WOODS" *and* *prefer natural to synthetic*

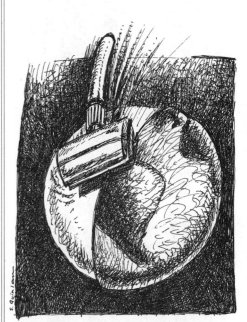

By the year 2000 it is expected that half the world's tropical forest will have been razed to supply the timber trade and make room for agriculture.

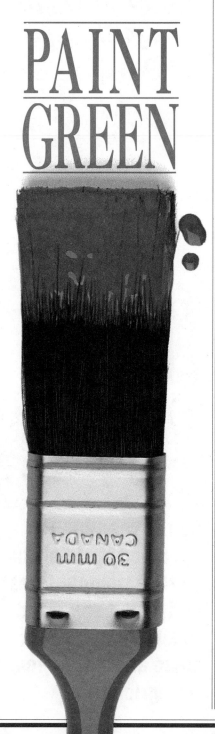

Veneers

You can assume that most tropical hardwood veneers come from unacceptable sources. Be wary of ebony, mahogany, African walnut, tulipwood, rosewood, and teak veneers (see above). Instead, ask for veneers made of apple, ash, aspen, beech, cherry, chestnut, elm, larch, maple, oak, pear, pine, poplar, or sycamore.

Carpets, Upholstery, and Drapes

For carpets, rugs, upholstery fabric, and drapes, stick to untreated wool, cotton, and linen. One reason is that synthetic fibres are based on petrochemicals and do not readily decompose when disposed of. Another is that synthetic carpets and upholstery may emit volatile chemicals — many new carpets, for instance, have a strong, distinctive odour thought to be caused by a component (4-phenyl cyclohexene) in the artifical rubber backing — that some people are sensitive to. The carpet industry recommends that carpet rolls be opened to allow odours to escape before installation, that alternative non-toxic adhesives be used to lay carpets, and that the area be well ventilated for several days after the carpet is installed.

The foam in cushions, mattresses, and carpet underlay, until very recently, was produced using CFCs. Furniture made with CFC-free foams is usually prominently tagged. Other acceptable upholstery stuffing materials are polyester, feathers, and cotton.

Paints and Solvents

Whether you should buy solvent-based or water-based paints is probably more a health than an environmental issue, and for most people, latex (water-based) is safer. Solvents (in strippers, oil-based paints, varnishes, thinners, sealers) contain chemicals that can be toxic enough that using them requires great care – adequate ventilation is a must.

A number of manufacturers now offer natural or non-toxic paints, sealers, strippers and other similar products. Some of these are listed in The Green Directory section at the end of this book.

A special word of caution: stripping paint can be one of the most dangerous of all do-it-yourself jobs. You're dealing with not only various stripping methods, all of which are hazardous, but also the old paint, which stands a good chance of containing lead if your home was built before 1950. If you can't avoid stripping the paint, be careful to use good ventilation.

And when you've finished any decorating job, dispose of all paints, varnishes, solvents, and strippers as hazardous wastes (see Chapter 7).

LEAD IN HOMES

While lead was eliminated from house paint in the 1970s, many older homes are coated with layer upon layer of lead-based paint. The best thing is to paint right over it. But if you plan to strip off old paint, you should have the lead content checked first. You can either chip off a small section and ship it to an environmental lab for testing, or call in an industrial hygienist to test the whole house with a X-RF (X-ray fluorescent) machine. Check in the Yellow Pages under "Laboratories."

If you are dealing with lead-based paint, it's safer to hire an expert firm that has the necessary personal protection and cleanup equipment to handle the removal job. If you decide to do it yourself, cover all surfaces with two layers of plastic sheeting and tape down the edges. Lead-based paints should be scraped off in large flakes; do not dry sand, burn off paint, or use a chemical stripper containing methylene chloride. Wear a high-quality dust mask and protective clothes (which should be laundered separately or, better yet, thrown away when the job is finished). Use a special HEPA (high-efficiency particle accumulator) vacuum to clean up; a regular vacuum cleaner will collect only the larger chips while spraying the finer ones around.

If you are removing leaded paints from the outside of the house, stil try to clean up all the chips and scrapings; lead does not break down in the soil and could pose a permanent toxic threat to humans and animals.

Lead in Drinking Water

Lead-free solder should be used to make any plumbing repairs. If you suspect that your drinking water has been contaminated by lead, you should:

- flush the contaminated water from your system by running the cold water tap for about three minutes every morning (if you take your shower first thing, you won't be wasting this water);

- keep a jug of cold drinking water in the refrigerator;

- use only water from the cold-water tap for drinking, cooking, or making up the baby's formula (hot water is likely to contain higher lead levels).

6. GARDENING

GREEN CONSUMERS

look for gardening methods and materials that produce tasty tomatoes and ravishing roses without hurting the environment or our own health.

▲▲▲▲▲▲▲▲▲▲▲▲▲▲▲

The same basic elements are needed to grow lawns, fruit, vegetables, or flowers: soil, seed, and care and protection.

Louis Dudek

"There is some hidden wisdom in all gardens."

▲▲▲▲▲▲▲▲▲▲▲▲▲▲

The soil in your garden, under your lawn and around your flowers, is a living organism. It takes up nutrients and excretes waste. The cycle of soil is the age-old tale of new life from old. Organic matter ages, dies, decomposes, and is turned into healthy young life. Last autumn's crimson maple leaves will, next year, be the green shoots of a new forest fern or wild flower. Green gardeners accelerate nature's cycle by composting.

Plants use organic matter to build their roots, stems, and leaves. In organic gardening, the soil provides some of the nutrients the plants need to grow. The soil is important because it contains bacterial organisms that break down the nutrients into a form the plant can use. When we apply compost, we are feeding the soil that will in turn feed the plants.

Aside from being harmful to the environment, commercial fertilizers may do your lawn and garden more harm than good. These fertilizers make grass and other plants grow much more quickly than they would otherwise, and disrupt the natural functioning of the soil and plant. Fertilizers give the soil a boost, but ultimately they weaken its constitution. The more chemicals that are used, the more disrupted is the natural process by which organic matter is converted into nutrients the plant can use. Eventually the soil dies.

In commercial use, as on large farm operations, rain washes a large amount of the fertilizer off the fields into streams and rivers, and then it is deposited in lakes. Water plants, including algae, thrive on the fertilizer and multiply rapidly. As they cover the water's surface, the surface plants block light and prevent it from reaching the underwater plants. In addition, as the surface plants grow and die, they use up oxygen in the water, thus depriving other marine life.

Fortunately, there's a simple solution at hand for home gardeners:

Composting

Like commercial fertilizers, compost supplies the three main nutrients needed for plant growth: nitrogen, phosphorus, and potassium. But it supplies them in a natural, "trickle release" system, so plants aren't burned by too much fertilizer, too fast; nor is the bulk of the fertilizer, like many commercial products, simply flushed away into the water table. You can have your soil tested (see pages 96-97) to find out what nutrients your garden particularly needs.

Liberal composting also dramatically improves the structure of the soil in your garden by improving both drainage and aeration. With each passing year of composting you will find the soil darker, richer, easier to work, and more free of pests.

These overall advantages of composting have been known for centuries. But in our polluted age, compost takes on special importance. Compost actually protects your plants from heavy metals in the soil and air, such as lead in the exhaust from passing traffic. Well-rotted compost "binds" or collects any lead in the soil, so that it cannot be taken up by your garden produce. It's a natural safety filter.

While it "imprisons" lead in the soil, compost releases other essential mineral nutrients to your plants. In addition to nitrogen, phosphorus, and potassium, compost has all the trace elements plants need – elements your garden seldom gets from commercial products. These include calcium, magnesium, sulphur, iron, manganese, zinc, copper, boron, molybdenum, and vanadium.

1 m

RAIN COVER

WOOD SIDES

EARTH

YARD WASTE

KITCHEN WASTE

BRICKS, STONES, OR TWO–BY–FOURS

YOU CAN MAKE YOUR COMPOST BIN YOURSELF

customized composting

Just as chemical fertilizers are designed with varying balances of nitrogen, phosphorus, and potassium, you can customize your compost to boost whichever nutrient is lacking in your garden soil. See below for a list of soil testing services that can tell you what help your soil needs.

	Nitrogen	Phosphorus	Potassium
beet roots	low	low	high
bone meal	high	high	low
coffee grounds	high	low	medium
corncobs (ground)	—	—	very high
eggshells	high	medium	low
feathers	very high	—	—
hair	very high	—	—
lobster shells	very high	high	—
oak leaves	medium	medium	low
peanut shells	medium	low	high
pine needles	low	low	low
pumpkin flesh	low	low	low
rhubarb stems	low	low	medium
seaweed	high	medium	very high

S·O·I·L TESTING SERVICES

Expert advice on what's in your soil now can help you to determine how to improve its quality. Soil testing services can tell you whether you need to boost the nitrogen, the phosphorus, or the potassium; whether your soil is too acid or too alkaline; and whether you have excess salinity. Some services can test for the presence of toxins and contaminants, such as lead.

Write to one of the services listed below, telling them what you want to find out. Their fees vary.

ALBERTA

Alberta Agriculture
Soils & Animals Nutrition Lab
9th Floor, 6909 – 116th St.
Edmonton, AB T6H 4P2

Alberta Environmental Centre
Plant Scientist Division
P.O. Bag 4000
Vegreville, AB T0B 4L0

CONTINUED NEXT PAGE 🖙

Compost Bins

You can make a compost bin yourself or buy one ready-made.

Your bin should be about 1 m square, at a minimum, and 1 to 1.6 m high. If it's smaller, it may not generate enough heat inside to decompose the material efficiently. If it's too much higher, the stuff on top may compress material on the bottom, and squeeze out the oxygen needed to keep the process cooking.

The sides can be wood, a circle of snow fencing, even wire fencing lined with cardboard cartons (to hold in the heat and moisture). If you use wood or cinder block, leave some air spaces or gaps. Put a few bricks, stones, or two-by-fours at the bottom to allow air to circulate underneath. Leave an opening in the front so you can turn the material over while it's cooking and remove the finished compost. Put a shed roof of wood or fiberglass sheeting on top to keep heavy rain off, or cover the top of the pile with plastic sheeting. If you can, build a double bin; you can continue to add to one side while the other is finishing off.

Ready-made compost bins are usually made of plastic, with a lid, and doors at the bottom to remove the finished compost. They are generally about 75 cm high and 60 cm square.

* *

Compost Pits

Compost "pits" work every bit as well as compost bins. So if you're a lousy carpenter, or don't demand geometric perfection in your garden, just dig a hole and throw your wet garbage into it. Cover it with leaves or grass clippings and a bit of the soil you removed. Dig some out from underneath when you want to use it.

Your compost pit should be kept moist. If there's a very heavy rain, and it gets too soggy, you could cover the pit with plastic until it dries out a bit.

* *

Making Compost

Start with a layer of coarse compost – small branches pruned from your shrubs and trees, straw, or grass cuttings, for example – on the bricks or loosely spaced wooden slats at the bottom of your bin. To get the process going well, add a layer of old compost, rotted manure, good garden soil, or seaweed, which is one of the richest sources of nutrients.

Now you are ready to add a layer of kitchen garbage. Any green kitchen wastes, such as vegetable tops and salad leavings, are terrific. Add coffee grounds and tea leaves, eggshells, fruit peels, and even scraps of paper (such as facial tissues or tea bags) if they're not too big.

Don't include plastic, glass, foil, or metal. If you don't want frequent visits from animals, skip fish and meat scraps, grease and oil, bones, eggs (including mayonnaise), and milk products, too.

When you add kitchen garbage to your compost pile, you're recovering a lot of nutrition that you used to waste. For instance, the average Canadian family throws away every year as much iron as is contained in 500 eggs, as much protein as they'd get from 60 steaks, and a volume of vitamins equal to the contents of 95 glasses of orange juice – just in discarded potato peelings. When you throw them into the bin or pit, you'll get the nutrients back in next year's backyard vegetables.

Now keep adding layers, with some soil, leaves, or grass cuttings between layers of kitchen garbage. If you are going to eat the garden produce next summer, don't use any leaves or grass cuttings that have been sprayed with herbicides or pesticides. If you run out of material, collect hair clippings from your local barber shop – no joke, they're rich in nitrogen. (Avoid, if you can, clippings with a lot of chemical colouring or conditioning on them.)

If you are in a hurry, and the weather is warm, you can have completely digested compost in two weeks by turning it with a pitchfork every two or three days, to keep it aerated and cooking evenly. If you just leave it alone, you'll have dry, crumbly, utterly inoffensive humus in eight weeks to six months. Poke at it and tumble it about occasionally, when adding new layers.

The speed of decomposition will depend on how much moisture and air is getting into the bin. You can help by mixing coarse and fine materials. A heavy layer of leaves, for example, can bond together like thick, wet cardboard and slow down the process. Punch holes in them with a sharpened broomstick or, better yet, put them through a leaf shredder in the fall. (If grass and leaves are stored in paper bags, they can be tossed directly into the pile – the paper will bio-degrade, too.)

You can tell whether things are working properly by the temperature. Once the compost pile is well started, the temperature inside (about 25 to 30 cm down from the top) should rise from 40°C to between 60 and 70°C, as decomposition starts the compost "cooking." It's "done" when the temperature falls back to about 40 to 45°C. The compost pile will steam as it's finishing the process, giving off moisture and heat. It will be hot to the touch while cooking and just warm when it is done. (If you want a precise measuring tool, garden centres sell soil thermometers.)

Put the finished compost on your house plants, your garden plants, and, after shaking or pushing it through a 0.5-cm screen, onto your lawn. There is no better fertilizer or soil replenishment.

* *

Composting Tips

▬ If your compost bin smells, it's too wet. Turn it over and mix in some dry material.

▬ If the temperature fails to rise, you might have too little moisture; sprinkle it with the garden hose. Or there might not be enough nitrogen to start the cooking: add some bone meal, blood meal, seaweed, grass cuttings, or manure.

▬ Composting works through the winter, too. Keep adding your kitchen garbage. If the scraps freeze, they'll break down into good soil all the faster in the spring.

▬ There are commercial "accelerators" for compost, but they should be unnecessary if you have followed all the instructions above. As a last resort, a carton or two of fishing worms should do it.

Norwest Labs
9938 - 67 Avenue
Edmonton, AB T6E 0P5

BRITISH COLUMBIA

Griffin Laboratories
1875 Spall Road
Kelowna, BC V1Y 4R2

Norwest Labs
203 - 20771 Langley Bypass
Langley, BC V3A 5E8

MANITOBA

Manitoba Soil Testing
Room 262, Ellis Building
University of Manitoba
Winnipeg, MB R3T 2N2

NEW BRUNSWICK

New Brunswick Department of
 Agriculture
Provincial Agricultural Lab
P.O. Box 6000
Fredericton, NB E3B 5H1

NEWFOUNDLAND

Department of Forestry & Agriculture
Soils Lab
P.O. Box 8700
St. John's, NF A1B 4J6

NOVA SCOTIA

Nova Scotia Department of
 Agriculture and Marketing
Plant Ind. Branch, Soil Testing Lab
P.O. Box 550
Truro, NS B2N 5E3

ONTARIO

Agrifood Labs
503 Imperial Road, Unit 1
Guelph, ON N1H 6T9

PRINCE EDWARD ISLAND

Soil and Feed Testing Lab
PEI Department of Agriculture
Box 1600
Charlottetown, PE C1A 7M3

QUEBEC

Soil Testing Lab
Macdonald Stewart Building
Room MS299, Macdonald College
Ste-Anne-de-Bellevue, PQ H9X 1C0

SASKATCHEWAN

Saskatchewan Soil Testing Lab
Department of Soil Science
General Purpose Building
University of Saskatchewan
Saskatoon, SK S7N 0W0

Mulch

Mulch is nature's compost system. You won't see much bare earth on a late-autumn stroll through any stretch of Canadian bush or forest. That carpet of leaves and needles is next year's soil nutrients – and this year's protection against wind and water erosion. It shelters the earthworms and other burrowing insects as they tunnel near the surface, aerating and draining forest soil and protecting tree roots from rot and exposure. As the birds and small animals poke at the mulch, winkling out insects and seeds, the matted leaves are broken up. Gradually they merge with the forest's topsoil.

Mulch is just as rich a bonus for your lawn and garden as for any forest. Providing mulch, or ground cover, for your garden has many benefits.

▬ Mulch smothers weeds, saving you the trouble of pulling them up. You'll have to take out the big ones, though – over 7.5 cm – before you put down the mulch.

▬ Mulch keeps water and nutrients in your soil by reducing evaporation, run-off, and wind erosion. It also adds more nutrients as it decomposes.

▬ By providing a layer of insulation, mulch reduces temperature shifts. The soil stays cooler in summer, warmer at night and in autumn. In winter, mulch reduces frost boils and breaks in the topsoil.

▬ With mulch you get earthworms, which provide aeration and drainage as well as the richest fertilizer of all: worm castings. Similarly, mulch encourages "good guy" bacteria and fungi, which nourish your plants, break down organic matter into rich new soil, and help repel pests or disease.

▬ Mulch protects the tender roots near the soil surface. It also keeps vine crops, such as cucumbers and melons, from rotting on the wet ground – and keeps them clean. Indeed, all your ripening vegetables will be cleaner, as the mulch prevents rain from splattering them with bits of soil.

▬ All the activity in and around mulch will help attract birds, foraging for stray seed, insects, and worms.

What are the drawbacks? Mulch can attract some unwanted visitors, too: slugs and mice.

To be rid of slugs, hand-pick them and move them to another part of greenery where they won't disturb you or your garden. Mice usually leave if you roll back the mulch carpet and make noise.

If you find your ground cover encourages crown rot in perennials, keep the mulch away from the base of perennials during their growth season. Don't let it rest too close to them in wet weather, either.

* * * * * * * * * * * * * * * * * * *

Mulch Ingredients

Availability, cost, and appearance are factors to consider when you select your mulch or ground cover.

▬ Grass clippings or leaves: The dry leaves should be shredded. Run them over with the lawn mower and put down a layer 5 to 8 cm deep.

▬ Seaweed: Terrific, if you can get it – it's full of useful minerals. Needs to be 5 to 8 cm deep. Apply after a rain, if you can, to avoid making the soil too salty.

▬ Sawdust or wood chips: Both leach nitrogen from the soil, so put a nitrogen-rich dressing down first. Both are high in carbon. Sawdust improves soil structure when it breaks down. Make certain it's not from chemically treated lumber. Wood chips are pretty; let them weather a year before you use them. Make a layer 2 to 5 cm deep.

▬ Corn cobs: Mills will provide these free; some mills will crush or grind them for you, so that they biodegrade faster. They, too, steal nitrogen from the soil. Spread about 10 cm deep.

■ Straw and hay: The most common mulches. Straw depletes nitrogen. If you use hay, look for last year's; you'll be less likely to spread a lot of nuisance seeds. Needs to be 10 to 15 cm deep.

■ Peat moss: It's clean, tidy, and attractive, and it greatly helps the soil hold moisture. But it's expensive and very acidic. Peat moss has to be wetted down when applied or it may blow away. Put down 2 to 5 cm.

■ Paper products: Cardboard is functional, if not brightly decorated with printing. It will eventually biodegrade. Newspapers are a bad idea: the inks have metals and chemicals you don't want in your food.

■ Black plastic film: Plastic keeps the soil hot and wet, lasts a long time, and is cheap. It's not very attractive, though. Punch some holes in it to allow for air circulation.

For cool-weather plants such as peas, beets, radishes, and lettuce, apply mulch in early spring to keep the soil moist and cool. For plants that need warmth – tomatoes, peppers, eggplants – apply it after the first growth spurt.

* * * * * * * * * * * * * * * * * * *

Vermicomposting

Good news for apartment dwellers — you don't need a backyard to compost. Vermicomposting, composting with worms, is the perfect alternative. It does not require much room, is odourless, and is easy. Perhaps the biggest fear is that the worms will get out of their bedding and end up in yours. They won't. To survive, worms must be in a dark, moist environment. They are perfectly content to stay in their vermicomposting container.

The size of the container you use and the number of worms you need depends on how much waste you produce. Start out with fewer worms. If you have too many worms and not enough food, many of the worms will die of starvation. If there is lots of food, the worms will reproduce. A good ratio is half a kilogram of worms to a bin about the size of a Blue Box or orange crate.

Just about any container with a lid and drainage holes in the bottom will do: a pail, a plastic tray, a flowerpot. Be sure to soak any porous container, such as clay, in water for a couple of hours before using. This will ensure that moisture from the compost is not drawn into the walls of the container but remains in the compost where it is needed. Another thing you will need is a tray to set your container in.

As in traditional composting, start out with a layer of somewhat coarse materials such as leaves, straw, or grass clippings. Mix with water until moist but not wet, and spread evenly in the bottom of the bin. Next, add worms. Redworms are best and are available from worm farms or bait shops for around $25 for half a kilogram. Spread the worms with some soil over the bedding and put the lid on.

Now all you have to do is feed them. Use the composting guidelines listed earlier. Chopping larger pieces into smaller ones speeds up the composting process. If flies are a problem, carefully dig the wastes into the bedding and cover with a light layer of soil. Keep the compost moist but not wet, and in a cool, dark place. Your worms will do all the work of turning and mixing the compost.

When your compost is done, push the finished matter to one side of the container and add new bedding and food to the other side. The worms will move out of the old matter towards the new food within a week or so. The finished compost can then be removed and used.

If You Can't Compost . . .

Buy products that will spare your environment while they nourish your garden. Most garden centres now stock natural organic fertilizers including cow, sheep, and poultry manure (dehydrated is odourless), meal (alfalfa, bone, blood, cottonseed), fish emulsion, and seaweed. All are acceptable alternatives if you can't make your own compost.

As a last resort, commercial organic fertilizers are available at most gardening supply centres as well as hardware stores, supermarkets, and other chain stores. Always check the labels: the material that makes up the biggest part of the product is listed first. For specific product names, check The Green Directory at the back of this book, ask at your gardening centre, or write to Canadian Organic Gardeners (see page 105) for their source list of organic fertilizers.

TREES
are more than just a pretty garden's face:
trees produce oxygen. The more of them we all nurture, the healthier our planet.

They also, of course, provide you with a shady spot on scorching days. And they shelter the birds that will eat up the insects you'd just as soon do without.

IF YOU WANT A FULLY ROUNDED GARDEN

– a private ecosphere – put in a perennial patch to attract the birds and insects you need to keep pests down, give your garden variety and colour, and give those helpful creatures the sanctuary they need.
Select a space you don't need for other purposes (the larger the better) and seed it with perennials: daisies, cornflowers, poppies, chamomile, columbines, and astors. Once established, this meadow will seed itself; you needn't mow it, and it will need minimal cultivation and tending. Let it "go wild" – that's the whole point.

"In the natural world, diversity is the key to survival."   David Suzuki

Most seeds sold through catalogues and at gardening centres are *hybrid* seeds, bred by artificial techniques at seed houses. These engineered varieties have little or no resistance to insects and disease, so they are dependent on chemical fertilizers, herbicides, and pesticides for survival. Most do not reproduce well, so you must buy new seed every year.

* *

The dominance of hybrids thus ensures that tonnes of chemicals will continue to be introduced into the food chain and washed into the water supply. At the same time, natural seed varieties are disappearing – and with them, the natural capacity for resistance that they have carried for thousands of years. Some scientists believe the decline in a wide variety of seed stocks is a more serious global problem than the destruction of the world's rainforests. To take only one example of shrinking variety in seeds, there were more than 8,000 apple varieties catalogued in North America at the turn of the century; today fewer than 1,000 remain.

Green Consumers prefer untreated, open-pollinated, or heritage seed. *Untreated* seeds are simply those that haven't been coated with chemicals to preserve them or to act as pesticides. *Open-pollinated* seeds are those that can reproduce themselves. That means you can save seed from your vegetables or flowers and use it next year. *Heritage* seed is handed down from generation to generation. Many gardeners have joined heritage-seed groups to encourage the preservation of seed species and to exchange seeds, thus preserving diversity in our seed stocks.

For home gardeners, the goal is seed that is "region-specific" (appropriate to conditions in one's area) and that has built-in, natural resistance to pests and disease. Most of Canada's seed for farms and gardens is imported from multinationals, but the major companies in Canada do produce some of their own, with regional varieties.

Check The Green Directory for suppliers of untreated, organically raised, and region-specific seeds. Many seed companies sell and ship organically grown seedlings and plants, too. Health-food stores may carry non-hybrid and untreated seeds, as well as organic garden products.

You can get involved in producing heritage seeds in your own backyard. When you have a good crop of anything – or your neighbour does – save some seeds to try again. Maybe you've stumbled on just the right strain for your soil, your climate, your sunlight orientation in the garden. After collecting the seeds, dry them out somewhere dry and warm (but never above 35°C, or you'll bake and kill them). Seal them in airtight containers, and put them somewhere as cool and dark as possible. You can even freeze seeds; the cooler and darker the storage area, the better. Heritage Seed Program (see page 105) can give you more information.

Don't hoard all the seeds, though. Leave a few sunflowers, gone to seed, for the blue jays and the chickadees. They make great emergency rations in early winter.

care &
PROTECTION

To the conventional gardener, "care and protection" means pesticides and artificial fertilizers. Green Consumers can replace most of these chemicals with composting, mulching, weeding, and cultivation . . . and a different attitude.

Changing your attitude means reconsidering what you identify as a weed or a pest. What makes a dandelion a weed to some and a beautiful flower to others? In Europe, dandelions are accepted rather than smothered with pesticides. But North Americans have decided that the bothersome dandelion does not belong on our manicured lawns. It disrupts the uniformity — the consistent-carpet-like texture and appearance. We have come to believe that this consistency is desirable. But it certainly isn't natural. Remember, the greater the diversity in your garden, the healthier the soil, and the better the ecological balance.

Grass produces up to three times as much oxygen as those plants we typically refer to as weeds. However, before you start weeding of any kind, rethink what you consider a weed — and why. Once you have decided what absolutely must go, get rid of these plants by digging and pulling rather than spraying and choking. Long-handled tools allow you to pull out a weed, root and all, without getting hunched over on your hands and knees. Once a weed has been pulled, replacing it with a piece of sod will better ensure that another weed does not take its place.

Weeds can be considered indicators of your lawn's well-being. It is often the case that they grow in conditions where grass cannot. If the soil is too wet, too dry, or lacking in nutrients, weeds may spring up where you had expected grass. If you are determined to have a lawn of only grass, it is important to realize that it doesn't make sense to spray with chemical pesticides. Poisoning the weeds does nothing to change the conditions in which they are thriving. Instead, find out what your lawn needs. Why are the weeds growing? What conditions need to change?

* * * * * * * * * * * * * * * * *

Watering Lawns and Plants

If water shortages are not a problem in your area, soak your lawn once every week to ten days. That means a solid watering of two to four hours for each section. You've watered enough if it "slurps" when you walk on it afterwards.

If water is in short supply, talk to your gardening centre about various sprinkler or watering systems for your lawn. Different types vary in price and maintenance — but they save a lot of time and are very efficient at conserving water.

In addition, sprinkle your plants several times each summer to wash off the residue from smog and traffic. Soot and grime on leaves interfere with photosynthesis and reduce both growth and crop yields.

If conditions are very dry, use the "clay pot" system for your plants, trees, and shrubs. The Chinese have used this method for over two thousand years, and many dry areas around the world continue to rely upon it.

KEEP · YOUR · GARDEN · GREEN

Mowing

Proper mowing is vital to having a healthy lawn. Mow frequently and high, leaving the grass 5 to 8 cm long. The longer the top growth of the grass, the longer and deeper the root of the plant. The longer the root, the healthier and stronger the grass. Nutrients stored in the leaf tissue are lost when the grass is mown too low. As well, long top growth will shade — and kill — most weeds. After mowing, grass clippings should be left down as a mulch to return nitrogen to the lawn.

THE
CHINESE
CLAY POT
SYSTEM

IT HAS BEEN IN USE FOR
DECADES; RECENTLY
IT HAS BEEN IDENTIFIED
AS THE POSSIBLE CAUSE
OF NON-HODGKIN'S
LYMPHOMA, A CANCER
OF WHICH CANADIAN
CASES DIAGNOSED HAVE
DOUBLED SINCE 1950.

Dig a hole beside each fruit tree or shrub that you want to irrigate, or every 2 to 3 m along a row of plants that you wish to water. Place an unglazed earthenware pot or jug in each hole. (Use old flowerpots if they are porous. To test them, stand them in water for a while; if they become moist inside, they are porous.) If the pots have drainage holes in the bottoms, plug or caulk them. Fill them with water and cover them to prevent evaporation. You'll find you need to fill them only once every four to eight days.

＊＊＊＊＊＊＊＊＊＊＊＊＊＊＊＊＊＊＊

Pest Control

Here is another place where a change in attitude is required. The principles involved in controlling insect pests are a lot like those applied to controlling weeds. A natural balance is maintained in the soil and vegetation between "pests" and their predators. When we apply pesticides we upset this relationship.

Less than 1% of insects are harmful to plants. Most are beneficial. Become familiar with the life in your garden; know who the real "pests" are and when to expect them. This will allow you to keep them in check before they cause excessive damage. You will also avoid damaging the rest of your graden — and the environment. Like weeds, bugs may indicate problem conditions in your garden. Remedy the cause, not just the symptom.

There are biological means of combatting or preventing most garden hazards, without resorting to chemicals. Hand-pick insects early in the season before they have a chance to reproduce; rotate vegetable crops as much as possible, planting disease-resistant varieties, and adjust planting times to avoid pests' active breeding periods; try companion planting – grouping together species that benefit each other while reducing the chance of

pest infestation or disease. Besides, not enough is known about the health hazards of most common pesticides and herbicides.

What we do know is that several chemical garden products contain possible cancer-causing agents. One extremely common weedkiller called 2,4D has been used for decades. It has been identified as a possible cause of non-Hodgkin's lymphoma, a cancer that is fatal in half to two-thirds of cases.

2,4-D is the most popular weedkiller in parks, on golf courses, and on suburban lawns. Some municipalities have stopped using 2,4D on their properties. Beware: some chemical fertilizers have 2,4D blended into them.

When we use pesticides, 60 to 90% (by volume) of what we spray misses the intended targets entirely. Most goes directly into our air or our water table, and then into our kitchen taps. We have little understanding of what effects result when the various pesticides combine with one another in the environment.

＊＊＊＊＊＊＊＊＊＊＊＊＊＊＊＊＊＊＊

Birds

In an ecosystem, poison for one can mean poison for all. Birds eat thousands of insects every day. They are a natural control of insect populations. Therefore, instead of killing off insect pests, look for ways to entice birds into your garden. Building a birdhouse, setting up a bird bath or planting flowers, trees, and shrubs with seeds and berries will all encourage birds to frequent your yard.

However, too many birds are occasionally a nuisance themselves — for example, pecking at the fruit in your backyard orchard. Scarecrows, foil pie plates, streamers, and wind chimes are often effective, but if stronger measures are required:

■ Buy some netting (with a 1-cm weave) from your garden centre, and cover the berry bushes or trees you want to protect. (The birds are afraid of getting their feet caught in the webbing.)

■ Buy a fake owl from your gardening centre. Perch it on a conspicuous branch. But move it every few days – birds are not as brainless as they seem.

✳✳✳✳✳✳✳✳✳✳✳✳✳✳✳✳✳✳✳✳✳✳✳✳✳✳✳✳✳✳✳✳✳✳✳✳✳

Killer Insects

Lots of insects eat other insects. Ladybugs eat aphids, mealy bugs, plant lice, and mites; dragonflies feed on mosquitoes; spiders, praying mantises, and wasps feast on many insects. A good number of insects can be purchased by mail. See The Green Directory at the back of the book. Insects, like gardeners, specialize, and are even finicky about their diets. Before you order an army, ask for a recommendation of the best warrior bugs to set on your whitefly, spider mites, aphids, thrips, or whatever insects you need rounded up.

✳✳✳✳✳✳✳✳✳✳✳✳✳✳✳✳✳✳✳✳✳✳✳✳✳✳✳✳✳✳✳✳✳✳✳✳✳

Do-It-Yourself Remedies

Tools and ingredients from the kitchen and the potting shed can be put to work against unwanted insects.

■ Slugs love to congregate under a board on the ground. Leave a couple in the garden. Flip the boards over every morning and scoop up the slugs.

■ Snails seek shade, so they'll be delighted to shelter inside a cool, shady clay flowerpot turned upside down. Gather them in the early evening.

■ Other insects can simply be hosed off sturdy plants — or wiped off with warm, soapy water. If you use the latter, rinse the leaves with the hose afterwards.

■ To post "keep out" signs for ants, sprinkle a line of paprika, red chili powder, dried peppermint, lemon oil or peel, or cream of tartar across their entrance.

■ Vegetable gardens can be protected from nighttime assaults by rabbits and groundhogs by planting rows of garlic and chives around them.

■ For a natural pesticide spray, throw some garlic and green onion tops into your blender or food processor. Strain this purée and mix the juice with soapy water. Pour into an old pump spray container and spray on leaves and stems of plants. Be sure to label the container!

Even when you use compost,

you'll still need to take measures against airborne lead that is deposited on the surface of your garden foods.

Fortunately lead is not absorbed through the skins of vegetables and fruit. But if they're grown within 100 m of busy traffic, you should wash them very carefully before serving them. Water alone won't remove all the lead deposits. Add a little vinegar or dish detergent to the water; scrub the food thoroughly, and rinse.

A windbreak – a fence or hedge – can greatly reduce the lead fall-out onto your garden from passing traffic. Just don't use the hedge cuttings in your compost for next year.

SLUG "PUREE" KEEPS THE SLUGS AWAY.

*If life gives you
dandelions —
make wine!*

Dandelions are high in iron,
potassium, and vitamins A and C.
The young leaves make wonderful
salad greens too.

Safer Pesticides

If all else fails, the products listed here are the least dangerous pesticides commercially available.

■■ *Bacillus thuringiensis:* BT is as safe as any pesticide can be. It will kill cabbage worms, cutworms, gypsy moth, and all caterpillars, including tent caterpillars. It's also been used successfully to destroy potato beetles and the larvae of black fly and mosquito.

You must apply BT directly to the plant or the soil (in the case of cutworms). Sunlight breaks it down, so use it in the evening. Rain washes it away; so applications must be repeated if it rains.

BT is sold under many trade names, including *Botanix, Dipel, Thuricide,* and *Envirobac.* Many garden centres sell it in boxes marked "organic garden spray."

■■ *Diatomaceous earth:* Not "earth" at all, this product is made of natural skeletons, like coral, whose crushed remains consist entirely of very sharp splinters. The splinters punch holes in the waxy shells of insects, so they die of dehydration. However, diatomaceous earth is not discriminating: it kills all the insects that encounter it. So it's probably best used indoors – where you don't want even beneficial insects – against earwigs, silverfish, ants, and cockroaches.

■■ *Rotenone and pyrethrum:* Both are "broad spectrum" insecticides (that is, they kill all the insects in their path). Extracted from plants, they are considered non-poisonous to people and pets when used as directed. Both will quickly despatch any cold-blooded animals, however, including frogs and fish, so don't use them near waterways.

■■ *Dormant oil spray:* Used by commercial orchards, it suffocates mites, scales, and other insects if applied to fruit trees before budding.

■■ *Tanglefoot:* This well-named product protects trees from caterpillars, ants, canker worms, and other climbers. One paints it around a tree, and they get stuck in it, so can't climb any higher – or lower for that matter. But it shrinks as it dries, so to avoid strangling the tree, paint a swatch of white latex around the trunk first. Then paint the Tanglefoot over the latex. You'll find Tanglefoot at most any gardening centre.

For More Information

In addition to the organizations listed below, ministries and departments of agriculture across the country will happily send you literature on composting and other features of organic gardening.

* * * * * * * * * * * * * * * * *

Canadian Organic Growers
Box 6408, Station J
Ottawa, ON K2A 3Y6

COG has a nationwide network of members and offers excellent information services. It also publishes a very informative quarterly newsmagazine on organic gardening, called *COGNITION*. Membership is available in COG and includes a subscription.

* * * * * * * * * * * * * * * * *

Civic Garden Centre of Metropolitan Toronto
77 Lawrence Ave. E.
Don Mills, ON M3C 1P2

The Civic Garden Centre operates a Master Gardener's Hot Line (416-445-1552) seven days a week from noon until 3 p.m. They can provide information on various types of gardening methods. They also welcome walk-in enquiries during the hotline hours in Edwards Gardens.

* * * * * * * * * * * * * * * * *

Ecological Agricultural Staff
Box 225, Macdonald College,
Ste-Anne-de-Bellevue, PQ
H9X 1C0

* * * * * * * * * * * * * * * * *

Friends of the Earth
251 Laurier Ave., Suite 701
Ottawa, ON K1P 5J6

Friends of the Earth has published a book on environment-friendly gardening. Called *How to Get Your Lawn & Garden Off Drugs*, the book is being sold in bookstores and directly by Friends of the Earth.

* * * * * * * * * * * * * * * * *

GROW
National Farmers Union
130 Slater St., Suite 750
Ottawa, ON K1P 6E2

GROW — "Genetic Resources for Our World" — is a clearinghouse and lobby formed to help preserve plant genetic variety, and to fight plant-breeding patent laws. Its members include the Canadian Labour Congress, the Canadian Environmental Law Association, Friends of the Earth, the United Church of Canada, and Oxfam-Canada.

* * * * * * * * * * * * * * * * *

Heritage Seed Program
c/o Heather Apple
R.R. 3
Uxbridge, ON L0C 1K0

The Heritage Seed Program is dedicated to saving unique seed varieties, fighting to preserve seed variety, and exchanging data. They publish the *Heritage Seed Program Magazine* and various booklets on how to save your own fruit and vegetable seeds. They have an excellent reference library and information base. A list is sent to members once a year from which they can choose free seed.

* * * * * * * * * * * * * * * * *

The Recycling Council of Ontario
489 College St., Suite 504
Toronto, ON M6G 1A5

Canadian Organic Growers highly recommends the Council's free booklet *Be Good to Your Garden, Compost! Your Guide to Backyard Composting*.

* * * * * * * * * * * * * * * * *

The Seed Savers Exchange
Route 3, Box 239
Decorah, IA 52101

The Seed Savers, a "mothership" for Canadians and Americans eager to preserve diversity of seed strains, publishes an annual list of over 4,000 "heirloom and endangered seed species." (Members receive this free.) Founded and operated by Ken Wheatly, Seed Savers publishes the *Garden Seed Inventory* each year, which lists all the Canadian and U.S. seed catalogues that include non-hybrid vegetable seeds. They also publish the *Fruit, Berry and Nut Inventory* list. Seed Savers operates an information clearinghouse.

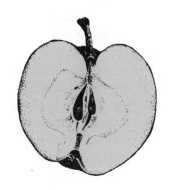

7. WASTE MANAGEMENT

THOSE OF US WHO LIVE IN CITIES

and towns rarely have the opportunity to realize just how much garbage each of us produces. It's whisked away regularly by our municipalities, and our surroundings become neat and ordered again. Only during garbage strikes do we begin to get some sense of the amount we throw away. And then it's only a few weeks' worth at most.

▲▲▲▲▲▲▲▲▲▲▲▲▲▲▲▲

If somehow we had an entire year's castoffs dumped back on our doorsteps, they probably would overflow the living room, creep into the kitchen, and possibly need bathroom space as well. There would sit our discarded bags, wrappings, food leavings, cans, bottles, junk mail, shoes, cosmetics, ornaments, papers, broken glass, appliances (large and small), cleaners, boxes, old dishes, razors, weedkillers, cardboard, furniture

(soft and hard), disinfectants, batteries, grass clippings, magazines, clothes, toys, paint strippers, medicines, suitcases, wood preservatives, tools, diapers, cat litter, jewellery, posters, games, umbrellas – all leftovers of our lives.

Until recently, most of us never worried about our mountains of trash. It's a big country, after all; surely there will always be space in which to hide our garbage. Trouble is, apart from the very serious environmental impact of landfill sites, getting rid of Canada's waste costs a great deal of money. And it isn't "government" money, of course – it's our money.

The garbage crisis of which we're becoming ever more aware is not really a problem of waste management. What we should be concerned about is our mismanagement of used materials and the fact that we produce garbage at all. But before we tackle those problems and offer some alternatives, let's consider how we handle our garbage now.

"Whatever befalls the earth, befalls the sons of the earth. Man did not weave the web of life; he is merely a strand on it. Whatever he does to the web, he does to himself. Continue to contaminate your bed and you will, one night, suffocate in your own waste."

Seattle, Chief of the Dwamish (1851)

Incineration

Burning our garbage seems like a good idea to many Canadians. Some even argue that we solve two environmental problems at once. We can get rid of our garbage while reducing the need for new coal- or nuclear-powered generating stations.

Unfortunately, this is not an environmentally sound solution; chemicals and waste don't simply disappear. About 20% to 30% of the wastes sent to the incinerator end up as ash or slag, while the remainder goes up the smokestack as fine particulate, water vapour, and gases. The solid wastes, which may contain a number of heavy metals and other persistent contaminants, must be disposed of in secure chemical landfills. And the atmospheric emissions, even after their passage through scrubbers and other stack emission controls, are by no means environmentally safe. The gas is primarily carbon dioxide, the main contributor to the greenhouse effect. Other air emissions include trace levels of toxic organic chemicals (dioxins and furans), heavy metals (cadmium, mercury, lead, etc.), and acid gases. Dioxins and furans are considered by some to be the most toxic of the organic chemicals, and incinerators are thought to be one of the major generators of dioxin.

Incinerators produce a wide range of other hydrocarbons that are harmful to humans and the environment. These include polyaromatic hydrocarbons (PAHs), chlorobenzenes, and PCBs. Although hexachlorobenzene is less toxic than the dioxins, it is emitted in far larger quantities, and very little information has been gathered on the effects of those emissions.

Mercury is just one of the heavy metals that gets into our atmosphere from incinerators. In Sweden, the government has calculated that 55% of all the mercury in the country's environment comes from incinerators. And mercury is very difficult to reduce because it's discharged in gaseous form; other metals are bound to fly-ash particles.

The acid gases produced by incinerators — sulphur dioxide, nitrogen oxides, and hydrogen chloride — have raised concerns for many years. In fact, incinerators are the single largest source of hydrogen chloride emissions into the environment. Unlike the other two acid gases, which can drift for hundreds of kilometres in the upper atmosphere before falling as acid rain, hydrogen chloride descends to earth rather quickly and may pose a health risk to anyone living within about 50 km of the plant. Emissions of sulphur dioxide and nitrogen oxides from incinerators are low in comparison to smelters, coal-fired generating plants, and other sources, but still add to the problem of acid precipitation.

Mercury from incinerators is believed to contribute to forest decline, and it is almost certainly the reason why fish in many areas of the world are accumulating mercury at unsafe levels.

The potential
energy savings
from recycling
paper, metals,
glass, and other
materials is far
greater than the
amount of energy
we get from
incinerators.

****

Energy from Garbage?

Even if we could control the air-emission problems, incinerators are still a highly inefficient way of producing energy. Because our garbage contains such non-combustibles as metal, a tremendous amount of the incinerator's heat is wasted simply heating up the metals. Energy must be used to boil off all the moisture in food, glass, and similar materials, too. In the end, only about 40% of the energy we get by burning used materials can be recovered to make steam: only about 15% can be recovered to make electricity.

In fact, steam and electricity produced by incinerators are created so inefficiently that we actually lose energy by burning garbage. The potential energy savings from recycling paper, metals, glass, and other materials is far greater than the amount of energy we get from incinerators.

****

Landfilling

Obviously, incineration of mixed garbage is not the solution to our waste-disposal problems. Why, then, can't we just bury it?

Landfills are no solution, either. For one thing, the sites chosen for them are usually on agricultural land, so landfilling contributes to the shrinkage of farmland. It's estimated that Ontario alone loses 0.6 hectares of land to garbage every day.

For another thing, no thinking person wants a landfill anywhere near his or her home. Landfills can leak. Their seeping liquid leachates can contaminate groundwater and surface water supplies. They give off greenhouse gases and other toxic airborne emissions, including methane, which is produced as garbage rots. Methane, in particular, is explosive; it can migrate from old landfills into the basements of houses.

Finally, the availability of cheap landfilling options can discourage waste recycling and reuse programs.

The question remains: what are we to do with the waste residues remaining after recycling and reuse have cut the waste load? The landfill sites of the future will have to be more environmentally secure than those we are familiar with — perhaps they will serve as huge temporary storage sites where we can securely store our used materials until better recycling/reuse technologies are developed.

Tomorrow's landfills will have to be sited in areas over clay soils (or other low-drainage overburdens) and isolated from sensitive land uses and sources of drinking water. They will need to be constructed with single or double liner systems, leachate collection systems, and monitoring wells. Strict controls must be placed on the wastes they can accept, and disposal fees charged sufficient to cover costs. Comprehensive post-closure/perpetual care plans will have to be drafted, and the methane gas collected (and burned to produce energy, or cleaned and sold).

solving the garbage problem

The solution to our garbage problem — our used materials mismanagement — is straightforward: stop creating garbage that must be disposed of and find ways to make the best use of the materials we do produce. In short, stop wasting resources and start conserving them.

A tin can doesn't become garbage until you throw it away. Up to that point, it was a highly refined and processed piece of tin and steel, one that was produced at an environmental cost (mining, energy for smelting, water for industrial processing, and so on). Turning some metal-bearing rock into a throwaway container is not only costly to the environment, it will become increasingly difficult to repeat as our resources run out.

Given that we can't replace that tin can effortlessly, it remains a valuable object even when it has been used. Its value, indeed, is equal to the total resources, capital, labour, and pollution expenditures that would be needed to replace it. It takes two to three times as much energy to make a new can as it does to recycle one. If we simply "burn" the can and turn it into incinerator ash and emissions, we squander all that energy. There are energy savings to be made from all recyclable materials, sometimes huge savings. Recycling plastics and aluminum, for instance, uses only 5% to 10% as much energy as manufacturing new plastic or smelting aluminum.

Long before most of us even noticed what we now call "the environment," Buckminster Fuller said, "Pollution is nothing but the resources we are not harvesting. We allow them to disperse because we've been ignorant of their value." To take one example, let's compare the throwaway economy with a recycling economy as we feed a cat for life.

Say your cat weighs 5 kg and eats one can of food each day. Each empty tin of its food weighs 40 g. In a throwaway economy, you would discard 5,475 cans over the cat's 15-year lifetime, not counting leap years. That's 219 kg of steel – more than a fifth of a tonne and more than 40 times the cat's weight.

In a recycling economy, we would make one set of 100 cans to start with, then replace them over and over again with recycled cans. Since almost 3% of the metal is lost during reprocessing, we'd have to make an extra 10 cans each year. But in all, only 150 tins will be used up over the cat's lifetime — and we'll still have 10 left over for the next cat.

Instead of using up 219 kg of steel, we've used only 6 kg. And because the process of recycling steel is less polluting than making new steel, we've also achieved the following significant savings: in energy use — 47-74%; in air pollution — 85%; in water pollution — 35%; in water use — 40%.

A tin can doesn't become garbage until you throw it away.

A cat may use up to 40 times its body weight in steel in 15 years.

So much unnecessary pollution, lost resources, and wasted land for garbage dumps happens as the result of a simple decision repeated again and again: we decide to ignore the valuable material left over after each feeding and call it garbage. That decision compounded in every aspect of our home life results in 10 million tonnes of household garbage per year in Canada, and about that amount again of garbage from our workplaces. This does not include hazardous waste, liquid waste and sewage, bio-medical waste, and air pollution emissions.

Between where we live and where we work, then, we throw out 20 million tonnes of used resources per year. That's almost a tonne for every man, woman, and child in Canada. A cat may use up 40 times its body weight in steel in 15 years; we throw out 150 times our average body weight in garbage during the same length of time.

You might expect that plastic would be a large part of our trash. But as you'll note, it accounts for only 5% – paper accounts for 36%. And our yard waste and food scraps account for more than a third.

Let's say you rethink your attitude to used materials. No longer is an empty glass, metal, or plastic container garbage; now it is a resource that can be recycled. No longer are grass clippings, leaves, and vegetable peelings garbage, they're a valuable mix of nutrients that can be composted and returned to the soil. Paper, too, can be recycled. That leaves you with the following:

Some plastic	3%
Composite materials	4%
Disposable diapers	2%
Textiles	1%
Miscellaneous	1%

If you also start using cloth diapers instead of disposables (see Chapter 4),

** *

Cutting Down on Garbage

If you have the average household (and of course few do), your garbage would break down this way.

Wet Waste	% by weight
Yard and lawn waste	25
Food scraps	10
Paper	
Newspapers	15
Mixed paper	10
Corrugated cardboard	3
Fine paper	2
Box board	1
Other paper	5
Glass	9
Metals	5
Plastic	5
Composite materials (two or more materials fastened together, as in blister packs)	4
Wood	2
Disposable diapers	2
Textiles (clothing, etc.)	1
Miscellaneous (hazardous household wastes, kitty litter, ceramics, etc.)	1

avoiding bad packaging, using hazardous waste depots, sending old clothing and furniture to charitable organizations, renting the items you use rarely and repairing others, you could cut the list of things you throw away to almost nothing. All that would remain would be odds and ends like lightbulbs (the glass of which can't be recycled), a few food scraps that can't be composted, food-stained paper (unless you composted it), and items that could no longer be repaired.

It can be done. In Seattle, Washington, where households are charged a certain rate for each garbage can they put out for collection, some homes have been granted a no-fee status because they produce no garbage!

It can't be done, of course, unless your municipality provides recycling services, your supermarket provides good packaging, and manufacturers make longer-lasting products. (Beware, for example, of throwaway cameras and disposable razors – especially the new electric ones.) If you've never even thought about your garbage before, you can get started right away by composting at home (see Chapter 6), by choosing refillable soft-drink containers, and by following the three Rs of waste management.

Wherever possible:

1 Reduce

2 Reuse

3 Recycle

These three are known as the waste-management hierachy – reduction is better than reuse, and reuse is better than recycling.

****

Reduce

R eduction is the most effective of the three Rs — you won't have to throw away what you didn't buy in the first place. Reduction means avoiding the purchase of gimmicks, wasteful products, and useless packaging. It means opting for good-quality, long-lasting products. And it means not buying what you don't need.

In the supermarket, for instance, are you going to buy pre-wrapped produce and carry it home in yet another plastic bag? Or are you going to choose loose produce and take it home in your own shopping bag? This point is important — about one-third of household garbage is made up of waste packaging. More money is spent by consumers on packaging than farmers receive for growing the food!

Other ways to reduce waste:
- Use cotton diapers instead of disposables.
- Use cloth napkins instead of paper ones.
- Don't buy disposable razors, pens, lighters, cameras, etc.
- Take your own refillable mug to work or school.
- Buy rechargeable batteries and a recharger unit.

Garbage Disposals

A garbage disposal — a machine installed just below the sink that chops food waste into tiny particles, liquefies it with cold water, and flushes it down the drain — wastes water and electricity and puts an increased load on the sewage treatment system.

Their supporters argue that using a "garberator" is more hygienic than trucking wet kitchen waste off to the dump, that they help extend a landfill's lifespan, and that a modern sewage works is perfectly capable of treating this kind of organic waste.

However, most environmentalists believe compost heaps are much friendlier to Mother Earth. While the sludge generated by a sewage treatment plant is difficult to dispose of (contaminated as it is with industrial wastes and heavy metals), the rich humus produced by a composter is a welcome addition to your garden.

Reuse

This R is often confused with recycling, but it means something different and is even better environmentally. Simply put, reuse is the act of making the same item serve over and over again. This sounds very basic, yet many of us miss a lot of reuse opportunities that would benefit the environment. Most of us have good intentions; we reuse yogurt containers for food storage and plastic grocery bags for carrying lunches or putting out garbage. How many people do you know with a drawerful of old plastic bags and a shelf laden with empty yogurt containers?

The other chances we miss include renting and repair. Rentals go far beyond cars, videos, and office equipment; just take a look at the Rentals section of your Yellow Pages. If you don't find what you want there, look under the item you're interested in, because just about anything for the home, garden, or repair shop can be rented. Whether you look at tools or tuxedos, society gets more value per item when consumers rent. One item is made, then used or worn many times by different people, and eventually retired after a job well done. Instead many tuxedos – and even more tools – are bought and used only a few times before being discarded.

Other ways to promote reuse include:

• Using repair shops to bring new life to worn but otherwise serviceable goods and appliances — resole shoes, reupholster furniture, repair broken televisions and radios. (And it keeps them out of the local landfill.)

• Selling unwanted goods through garage sales, auctions, classified ads, bulletin board ads, secondhand dealers and antique stores.

• Donating unwanted goods to charities, which are always looking for clothes, bicycles, toys, books, magazines, appliances, furniture, and all sorts of other things.

• Buying foods and other products from bulk dispensers, allowing you to reuse containers again and again. If we can return beer and pop bottles for reuse, we should be able to carry a clean, empty container to a food store for a refill.

****

Recycle

What makes recycling different from reuse is that the item involved must be reprocessed — an extra step that may have some effect on the environment. Paper gets repulped and made into new paper, but obviously the processing results in some pollution.

However, the savings to be gained from recycling are considerable. For instance, 2.2 tonnes of wood are needed to make a tonne of new paper, along with considerable quantities of energy, water, and chemicals. Making paper from recycled fibre uses 40% less energy. Glass can be recycled forever, and every tonne of crushed waste glass used saves 135 L of oil and 1.2 tonnes of raw materials. A tonne of aluminum uses up 4 tonnes of bauxite, and all metal smelting is energy-intensive. The energy used to make one tonne of virgin aluminum could recycle 20 tonnes of aluminum from scrap.

Here is just a brief list of what can and what can't be recycled in some parts of Canada:

YES
(in your Blue Box)

- Newspaper
- Corrugated cardboard
- Glass bottles and jars (rinse, and remove all caps and lids)
- Aluminum and tin/steel cans
- Plastic bottles and jugs (including those used for soft drinks, detergents, juice, bleach, water, shampoo, motor oil, etc.)

YES
(Using Private Companies, Charity Groups, or Special Municipal Programs)

- Fine paper
- Computer paper
- Some telephone books
- Eye-glasses
- Used motor oil
- Car batteries
- Coat hangers (accepted by many cleaners)
- Large appliances
- Automobiles
- Steel, copper, and most other metals
- Some rubber tires
- Food wastes (in backyard or communal composters)
- Leaves, trimmings, and other yard waste
- Christmas trees

NO
(at least, not yet)

- Aerosol cans
- Disposable diapers
- Composite materials (paper and/or plastic and/or metal foil joined together in packaging)
- Waxed and coated boxes, pizza boxes, cereal and shoe boxes
- Paint cans
- Most small batteries
- Glossy magazines
- China
- Clay pots and earthenware
- Porcelain (insulators, bath tubs, toilets)
- Crystal
- Mirrors and window glass
- Lightbulbs

Recycling Programs

The Blue Box program in Ontario is the largest and most successful recycling effort in North America. About 2 million households in the province have each been given a heavy-duty plastic box for regular pickup of cans, glass bottles, plastic soft-drink bottles,

Neunkirchen, a county in Austria, recycles two-thirds of all household and commercial/industrial used materials.

It's Only Recyclable if Someone's Recycling It

These days, it seems everyone is calling their product "Recyclable". It's true that almost any waste — disposable diapers, aerosol cans, Tetra-Pak drink boxes — can theoretically be recycled. But ONLY if there's the time, energy, and money to do the job. There has to be a collection and sorting system, a recycling facility in full-scale operation, and viable markets established for the recycled end-product. Otherwise, such claims are just hollow advertising slogans.

There is no question that we're an overpackaged country.

and newspapers. Some municipalities are also collecting corrugated cardboard (the type that has a ruffled layer sandwiched between two flat layers), used motor oil, and all hard plastic bottles. The programs are expanding gradually to include more households.

The Blue Box program came about as a result of an amendment to the Ontario soft-drink container law., Almost every province has long had regulations governing these containers, generally limiting the use of various types and stressing that a certain percentage must carry a deposit and a certain percentage must be recyclable.

The best example is Prince Edward Island, where all soft-drink containers must be refillable. The original purpose of these laws, most of which were drafted in the early 1970s, was to reduce the amount of roadside litter. It has only been in the past four or five years that governments have begun to see the implications for used-materials management inherent in their soft-drink container laws.

The success of the Blue Box program has helped convince other governments that recycling can work on a large scale and that even more can be done. Similar programs are springing up all over. British Columbia, Nova Scotia, New Brunswick, and Quebec are taking a serious look at the Blue Box idea. Each should have a number of programs operating by 1991. Some Alberta cities are already using the Blue Box.

Despite its success, the Blue Box hasn't solved the problem of used materials mismanagement. The best examples of the program — Ottawa, Peterborough, Mississauga, and Guelph — are getting only about 15% of the garbage from single-family homes. And that is only about 5% of the total garbage volume in those cities.

So environmental organizations are drawing the attention of Canadian governments to vastly more intensive efforts in Japan, West Germany, and Austria. The most successful program, in Neunkirchen, a county in Austria, recycles two-thirds of all household and commercial/industrial used materials. It's worth noting that Europe produces fewer and smaller packages than North America does, and Europeans tend to be less wasteful in what they buy. Austrians throw out only a third of the garbage we do; with two-thirds of that being recycled, their total waste is one-ninth of ours.

Many municipalities are considering the Austrian system, which relies on a simple separation of garbage. Each household receives two containers, one of them a sort of super Blue Box for dry used materials like cans, glass, and all paper. Its contents go to a separating plant, where metals are taken out by magnets, paper is separated out by air clarifiers (blowers) and gravity, and other materials are pulled out by other means.

The second bin is for wet garbage: food scraps, lawn clippings, and other yard waste. This material is collected and composted centrally, either in a plant or in large "windrows" out of doors. The compost is sold to plant nurseries and parks, and is acceptable for agricultural use under Austria's quite rigorous standards. But the preferred route is that individuals compost in their own backyards where possible; this saves collection and processing costs and produces a cleaner compost. Private industries can deliver their wastes to the sorting or composting plant for a fee that's lower than the cost of disposal elsewhere.

The village of Ryley, Alberta, already uses a similar system. Guelph, Ontario, has set up a pilot composting program designed to serve 600 households, and 35 other Ontario municipalities established leaf composting programs in 1989.

The Perils of Packaging

The production, use, and disposal of the materials we use to wrap our foods, clothes, and other consumer goods contribute to many environmental problems, from litter through waste disposal to acid rain.

The more expensive the item, the more layers of wrapping. Witness that expensive box of chocolates or the pricey mushrooms in a plastic box wrapped in plastic. Few products need more than one layer of packaging, almost none need more than two layers. Yet a common sight in supermarkets is two or four tomatoes sitting in a cardboard box and surrounded by plastic wrap.

Traditional packing materials have given way to foamed plastics and aluminum; plastic wrappers and the plastic rings on six-packs (banned in many U.S. states) strangle and suffocate birds and marine animals. Although paper and cardboard eventually break down as waste, they're rarely made from recycled material. In a year each of us might use two trees' worth of packaging material.

Even juice boxes, valued because they're light and store neatly without wasting space (hence saving fuel used in transportation), are environment-unfriendly because they're not being recycled. But instead of buying one large recyclable glass container of juice, we opt for boxes made from plastic wrap, cardboard, and foil, then packaged in sixes wrapped in more plastic. And biodegradable and photodegradable plastics aren't a solution either; many environmentalists feel these will only leave us with plastic dust, the environmental impact of which isn't known.

Consumers *do* have a say in packaging. Their reaction, for example, removed a new plastic soft-drink can from the market in the U.S. Some U.S. states are imposing waste taxes on packaging materials that can't be recycled, which are added to the price of products like toothpaste. There's also a shopping bag tax: bring your own bag to the store (and get a rebate, in some places) or pay for a new one.

A British borough council has a campaign called Don't Choose What You Can't Reuse. This won't do the whole job; reuse must be combined with recycling and — most important — reduced consumption. But every bit counts. Here are some measures you can take that will help considerably.

• Use your municipality's Blue Box recycling program, if it has one, for cans, bottles, and acceptable plastics.

• Choose returnable bottles.

• Don't litter anywhere with anything.

• Cut down on the amount of plastic and foam packaging you bring home.

• Buy products made from recycled materials; if they're not available, ask why.

• Reuse plastic bags when you can't avoid buying foods in them. For example, plastic milk bags are excellent for storing foods.

* *

Degradable Plastics

Many Canadians are concerned about the amount of plastic we add to our garbage – 16 to 18 kg per person per year. We know those plastics will pile up over the centuries as a lasting monument to our throwaway society. All of us have seen, too, the many plastic items that aren't put out in the garbage; they litter our roadsides, parks, and campgrounds. And so the new

Paper
vs.
Plastic Bags

There's been a lot of argument over who is really the better environmental citizen. Paper makers say their industry is based on a renewable resource, unlike the petroleum-based plastics industry. Plastic manufacturers insist that more air and water pollution is generated and more energy consumed making a paper bag than a plastic bag. And although paper is biodegradable and can be composted, it can take decades to decompose when buried deep in a modern landfill site. Of course, plastic bags take even longer to break down — they will be with us for centuries. Neither paper nor plastic shopping bags are currently recycled to any great extent.

Don't Use Either!

Instead of arguing which is the lesser of two evils, the Green Consumer leaves the bag behind in favour of a reusable tote bag. If you need to take a bag, choose one you can recycle or reuse later.

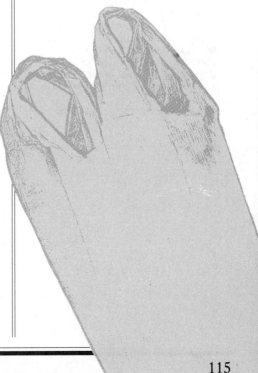

A Return to the Old Ways

Vancouver residents now have a special opportunity to reduce waste . It isn't anything new or fancy, but it's something most Canadians seem to have forgotten about. The era of milk deliveries has returned to Vancouver, and it has brought back the refillable glass milk bottle. During 1988, **Avalon Dairy Ltd.** sold 75,000 L of milk in returnable glass bottles.

Not many communities have dairies like Avalon, but plastic jugs for milk are fairly common. Those jugs can be recycled (if not refilled), but in fact most of them aren't.

Becker's, a jug-milk chain with more than 700 corner stores in southern Ontario, recycles its two-litre jugs when customers return them to the stores. And 98% of jugs do come back, largely because Becker's charges a deposit on each jug. To encourage buyers to choose the recyclable jugs rather than disposable milk bags, Becker's sells the jug milk at a lower price than the bags.

biodegradable and photodegradable plastic bags seem on the surface to be very beneficial. But we may have to think again about these wonder products.

Biodegradable plastics are made by including cornstarch with the polyethylene in the manufacturing process. When the plastic item is discarded, the cornstarch is eaten by bacteria, which causes the polyethylene to break down into pieces too small to see. Another chemical added during manufacture helps provide oxygen to the bacteria, which also aids in the decomposition of the plastic.

Photodegradable plastics are used in bags and in the plastic rings used to hold six-packs of cans together. In this case, a light-sensitive additive is mixed with the plastic to help break the product down into small particles in the presence of sunlight. This is important environmentally not only because of the despoiling of nature by litter, but also because seabirds often choke or strangle on plastics.

Plastics that will degrade in a few years, or even a few months, seem to be promising solutions. However, let's apply two of our rules for managing waste and see how they come out.

* *

Reduce, Reuse, and Recycle

RULE ONE

Do degradable plastics reduce the volume of garbage we generate? Not at all. In fact, in order to guarantee the strength of biodegradable bags, approximately 5% to 10% more plastic must be used in their manufacture.

Are degradable plastics reusable? Obviously not, since they're designed to fall apart.

As to whether they're recyclable, the answer on one level is yes. At least one company recycles the scrap and trimmings from manufacturing biodegradable bags into new bags. This recycling of in-house scrap is standard practice in the manufacture of regular plastic bags, but it is an expensive new technology in the biodegradable field. Although this may change in time, as of now it means that more of the scrap from these plastics will go to the dump.

On another level, the issue is even more complicated. In recent years we've started to get good news about the recycling of "post-consumer" plastics (meaning those you have already disposed of). A number of communities have begun to recycle some hard plastics such as soft-drink bottles, jugs, and magarine tubs; many other municipalities are expected to follow suit shortly. In Korea, West Germany, and some other countries, they are recycling film plastics, including plastic bags. And new technologies continue to be developed, including one that can fuse together all kinds of plastics and make a wood substitute for fence posts and park benches.

The problem is that degradable plastics may interfere with some of these recycling technologies. The degradables may be difficult to recycle themselves, and they make it difficult or impossible to recycle other plastics with which they are mixed. Plastics manufacture is an extremely precise business, and recyclers must meet exacting standards of purity. Degradable plastics will only complicate the process.

The key question with degradable plastics, then, is whether they make recycling easier or whether they hinder the process. Since they cause problems, it makes more sense to stick with the regular plastics in the hope that improvements in technology will make them fully recyclable.

THE ANSWER

So while these individual pieces of plastic may not be visible to our eyes, they may still be with us for centuries. Even if we replace all ordinary plastics with these special new plastics, we will still be collecting and transporting the same amount of discarded packages to landfill sites. And once they are buried, we will have wasted forever the resources used to make them. Whether they are dust or bags, they will have cost us dearly.

As with all other materials, the best route to take is still:

Reduce. Don't accept any bags you don't need — plastic, paper, or whatever. Use permanent shopping bags of sturdy cloth, canvas, or string.

Reuse. Use any bags you do collect again and again.

Recycle. Urge your municipality to adopt a plastic recycling program — and when you get it, use it faithfully.

THE DISPOSAL OF
HAZARDOUS
WASTE

As outlined in other chapters – on cleaning, the home, and gardening – many of the products we use in and around our homes pose a threat to the environment.

The best way to make sure these products cause no harm is to avoid purchasing them in the first place; substitute non-toxic or less hazardous products for the ones you are now using (see Chapter 3 for a number of non-toxic alternative cleaners you can make yourself). However, if you have already bought some of these hazardous products or found them gathering dust in storage areas in your home, it's vital that you dispose of them in the safest possible manner.

Never pour them down a drain or put them out with your garbage. Hazardous liquids can corrode plumbing, release toxic fumes, and damage septic tanks and sewer systems. And every year, sanitation workers are injured when containers of hazardous waste hidden in the trash unexpectedly rupture. Workers at sewage treatment plants have also complained of health problems thought to be linked to hazardous materials dumped in the sewers.

Moreover, municipal landfills and sewage treatment plants aren't designed to cope with hazardous materials, which may escape into the air or groundwater. Incinerators may just hasten these chemicals' entry into the environment.

Our garbage contains these potentially toxic materials:

Ammonia-based cleaners
Barbecue starter fluid
Batteries
Brake fluid
Butane lighters and cylinders
Chlorine bleach
Drain cleaners
Flea collars and sprays
Floor wax
Furniture polishes
Lighter fluid
Medicines
Metal polishes
Mothballs
Nail polish remover
Oven cleaners
Paints and paint thinners
Paint stripper
Pesticides
Photo-developing chemicals
Rug and upholstery cleaners
Solvents
Spot removers
Stains and finishes
Swimming pool chemicals
Toilet cleaners
Transmission fluid
Turpentine
Used motor oil
Window and glass cleaners
Wood preservatives

In many areas, you can take these products to a permanent household toxic waste depot set up by your provincial or municipal government, or to a temporary collection facility on a Special Waste Collection Day. Some wastes can be reused or recycled, others can be chemically neutralized. Some must be sent to properly licensed facilities that can dispose of them safely (through neutralization, high-temperature incineration or secure chemical landfilling).

The following list will tell you if your area has such a program for hazardous waste, and whom to contact. (You'll note that some provinces have no programs, and you might like to ask why.) If you don't have a car to get you to the waste depot, combine forces with a friend or neighbour who does have transportation. That way you can dispose of all the chemicals from two (or more) homes at once.

British Columbia

Special Wastes and Emergencies Unit, Waste Management Branch, Ministry of Environment, Victoria, (604) 387-9952. Nine regional offices operate depots: Vancouver Island, 758-3951; Cariboo, 398-4532; Lower Mainland, 584-8822; Skeena, 847-7260; Thompson-Nicola, 374-9717; Omineca-Peace (Prince George), 565-6135; Omineca-Peace (Fort St. John), 787-3295; Kootenay, 354-6121; and Okanagan, 493-8261.

Alberta

For the times and locations of Toxic Waste Roundup days, call the Alberta Special Waste Management Corporation (403) 422-5029 or (800) 272-8873. It will also help set up and for the time being fund community hazardous waste collection programs. Edmonton operates its own roundup for three days once a year.

Saskatchewan

Contact Manager, Chemical Management Section, (306) 787-6185. The City of Regina has been holding householder hazardous waste collection days for several years. An agricultural chemical collection program has also been established for rural areas (including mobile collection for farmers unable to transport).

Manitoba

The Manitoba Hazardous Waste Management Corporation and the City of Winnipeg operate a facility at 745 Logan Ave. every Saturday. Other communities may run programs in conjunction with MHWMC. Ask your municipality or call them at (204) 945-5781.

Ontario

Metropolitan Toronto runs four permanent depots (416) 392-4330 and a Toxic Taxi program. Depots also operate in Peel Region, (416) 566-1511; Halton Region, (416) 827-2151, ext. 470; and Waterloo, (519) 885-9500. The province has a cost-sharing program to encourage other municipalities to run single-day depots. For information, contact the Household Hazardous Waste Program, Waste Management Branch, Ministry of Environment, (416) 323-5202.

Quebec

The Montreal Urban Community Association (514) 280-4330 organizes collection days in that city.

New Brunswick

The province has no permanent depots but has run pilot programs in Fredericton and Bathurst region. It also operates a permanent fuel collection program. For more information, call (506) 453-3700

Nova Scotia

There has been a collection day in Halifax organized by the Ecology Action Centre; call (902) 422-4311.

Prince Edward Island

The province offers occasional one-day depots. Contact the Supervisor of Air Quality and Hazardous Waste, Department of Environment, (902) 368-5031.

Newfoundland

No program at present.

5 BAD PACKAGES

Until our various levels of government start to work together on reducing our packaging and hence our garbage, do your best to avoid the following examples of environment-unfriendly product wrappings.

1 Tetra-Paks: These are the square juice boxes you see in vast numbers on supermarket shelves. They can be neither reused nor recycled (to any appreciable extent) because they have an outer layer of plastic, a middle layer of cardboard, and an inner layer of aluminum foil that are difficult to separate. Some pilot recycling projects are under way.

2 Blister packages: These are cardboard-backed packages with a plastic bubble on the front to hold the product to let you see the contents. Again, however, the different materials can't be separated efficiently for recycling, and so they become general garbage.

3 Individually wrapped snacks: A package of cookies that has each cookie separately wrapped may seem "hygienic" at first glance, but is the second wrapping truly necessary? And do we need individually wrapped restaurant portions of butter, sugar, salt, pepper, ketchup, and other condiments?

4 Single-serving microwaveables: Here, too, the layers of packaging are excessive. We could easily buy a larger package, take out the appropriate serving size, and heat or cook it on a plate or pan.

5 Tea bags: The tea leaves come in an absorbent bag with a string attached, which is tucked inside a separate paper envelope, which is packed inside a cardboard box, which is wrapped in plastic. Loose tea in a tea "egg" is much more civilized, in every sense of the word.

8. TRANSPORTATION

IN CONQUERING CANADA'S VAST SPACE, *modern transportation has done nothing less than make our national life possible, but at considerable environmental cost. Second only to industry in its consumption of energy, transportation accounts for one-quarter of all energy used in the country, with road vehicles responsible for 83% of that share.*

▲▲▲▲▲▲▲▲▲▲▲▲▲▲▲

It's also the country's largest source of air pollution, every year spewing into the atmosphere some 13.6 million tonnes of noxious fumes that poison forests, lakes, and marine life, contribute to global warming, and endanger human health. According to Environment Canada, transportation sources produce 64% of total nitrogen oxides (a cause of acid rain), 42% of hydrocarbons and 66% of carbon monoxide (both of which cause smog), 32% of the lead, 30% of the carbon dioxide, 76% of the benzene (a carcinogen), and unknown

▲▲▲▲▲▲▲▲▲▲▲▲▲▲

quantities of toluene, xylene, and ethylene dibromide.

Transportation also takes up space. To build all the necessary highways, roads, docks, railway tracks, and airfields, millions of hectares of forest, farmland, and cityscape have been sacrificed to the bulldozer.

By far the biggest players in the scenario are gas-powered motor vehicles, especially the automobile. Cars and trucks are much more damaging to human and environmental life than any other form of transport. A Swiss study, for example, found that, compared with trains, motor vehicles accounted for almost three times as much land use, 3.5 (cars) and 8.7 (trucks) times the energy consumed, nine times the pollution, and 24 times the accident rate. (Canadian figures would probably be similar or even worse for motor vehicles.)

> "The car has become a secular sanctuary for the individual, his shrine to the self, his mobile Walden Pond."
>
> E.C. McDonagh

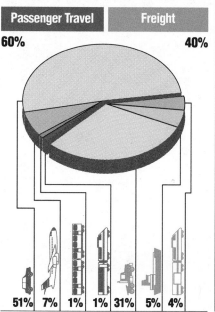

Transportation Uses of Energy (%)

Passenger Travel	Freight
60%	40%

51%	7%	1%	1%	31%	5%	4%

W e may all want to go back to nature, but we want to go back in a car. In the six decades since it became available to almost any North American over 16, the car has not only transformed the way we move about but assumed an almost sacred place in our lives and psyches. We have come to regard its gift of unprecedented individual mobility as a basic right. In 1986, Canadians owned 11.5 million cars, one for every 2.23 people. In global terms, the allure of the car has proven contagious: since 1950, the number of cars worldwide has jumped from 50 million to 386 million, and the total is continuing to rise. In the Soviet Union and Eastern Europe, car fleets jumped 500% from 1970 to 1985; and even in the developing countries the number of cars, though comparatively small, is growing at twice the rate as in the industrialized nations.

OUR MOTHER, THE CAR

Yet the car's very success has brought a Pandora's box of modern ills that would have appalled Henry Ford: air pollution, used resources, noise, congestion, destruction of land and wildlife, human death and injury – all on a grand scale.

The following are some of the particular issues the Green Consumer should consider in assessing the environmental effects of the car.

* *

Losing Ground

The environmental group Friends of the Earth estimates that each kilometre of road or highway takes up about 6.5 hectares of land. In Ontario alone there are 155,000 km of highways, roads, and streets, adding up to a million hectares of land given over to motor vehicles. Our cities devote one-third of their area to roads and streets. Highways consume the largest amount of space per kilometre, but even relatively small roads can cause major controversy where they run through environmentally sensitive areas.

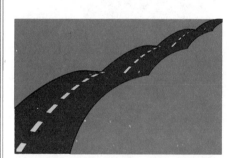

Each kilometre of road or highway takes up about 6.5 hectares of land.

* *

Fuel Efficiency

C onsumer pressure has been a key factor behind the trend to more fuel-efficient automobiles. Fifteen years ago, with the world still reeling from the shock of the first OPEC oil crisis, the automobile industry was slowly forced to recognize that it would have to junk its gas-guzzlers and produce thriftier cars. And it was successful: the average car's fuel efficiency improved from 22 L/100 km to 15 L/100 km, with much higher ratings achieved by some Japanese and Western European car manufacturers. U.S standards for 1990 model cars will require fuel efficiencies of 8.5 L/100 km; indeed, levels of 5 to 6 L/100 km should be easily achievable using current technology for only a small increase in car prices. For example, the 1990 Chevrolet Sprint (manual) is rated at 5.5 L/100 km (city) and 4.4 L/100 km (highway). Between 1972 and 1982, the total amount of vehicle fuels consumed in the leading car-producing nations fell by 4%, even though the number of cars in use jumped by a third.

1,500 L OF OIL

TO MANUFACTURE THE AVERAGE CAR

10,000 L OF GAS

FUEL USED UNTIL THE VEHICLE IS SCRAPPED

Yet we still need enormous quantities of fuel to keep us on the move:

■ Because other sectors have reduced their reliance on oil since the two OPEC crises, cars, which operate almost exclusively on petroleum-based fuels, actually account for a larger portion of oil use now (about a billion tonnes per year) than they did then.

■ World gas consumption has edged up again since 1982, thanks to cheaper prices and a false sense of oil security. The truth is, all known oil reserves are expected to be used up in about 30 years.

■ In Canada and the U.S., generally lower fuel-efficiency ratings and more hours on the road mean the average car burns up twice the amount of gas that its counterpart in Western Europe or Japan does.

■ Even today, the average car takes the energy equivalent of 1,500 L of oil to manufacture and uses at least 10,000 L of fuel before it's scrapped.

The fuel economy of new cars is now a key selling point, and designers have achieved a great deal in making cars more efficient. The car industry's challenge for the 1990s will be both to maintain and improve on these achievements and to design cars that are safer, cleaner, quieter, and longer-lasting. The challenge for the Green Consumer will be to help persuade the industry that cleaner, quieter, and longer-lasting cars are what the customer wants.

Pollution

As factories, power stations, and other industrial sources of air pollution begin to clean up their acts, motor vehicles in all industrialized nations have emerged as the worst offenders. The average car, for instance, pumps five or six times its weight in carbon into the atmosphere each year. Despite get-tough environmental regulations and technical improvements in fuel economy over the past 15 years, the volume of most pollutants produced continues to grow. Many of the positive effects are being offset by an increase in sheer numbers of cars and kilometres of driving.

Cars produce six major forms of pollution:

① **Hydrocarbons.** These gases react primarily with nitrogen oxides in the presence of sunlight to produce photochemical smog, the familiar urban haze that causes respiratory disorders and erodes buildings. Smog's principal component is ozone, a greenhouse gas and a suspected cause of forest damage. Gas-powered motor vehicles produce almost half of all hydrocarbon pollution in Canada.

② **Nitrogen oxides.** Produced by all combustion processes, from power stations to motor-scooters, nitrogen oxide emissions are implicated in smog and, with sulphur oxides and unburnt hydrocarbons, in acid rain.

③ **Carbon monoxide.** The gas that kills you if you leave your car running in a closed garage, carbon monoxide is actually increasing in volume in most industrialized countries. Environment Canada says it's increasing here, too, although the share accounted for by transportation sources is actually decreasing.

4 Particulates. Road vehicles, especially diesel-powered ones, have overtaken the coal fire as the major source of soot. Particulate matter reduces visibility, blackens buildings, damages plants, and is implicated in the onset of cancer and other diseases.

5 Lead. Among the most ubiquitous of pollutants, lead at sufficient blood levels damages the brain and central nervous system; children are particularly vulnerable. However, lead is the one exhaust emission for which there is real hope of a dramatic reduction, as stringent lead restrictions are being enforced in many countries, including Canada.

6 Carbon dioxide. Although carbon dioxide is not a regulated automotive emission, as a greenhouse gas it is associated with perhaps the most serious long-term environmental threat of all, global warming.

What is being done – and what could be done – to control these pollutants?

* * * * * * * * * * * * * * * * *

Getting the Lead Out

G asoline is a blend of up to 400 hydrocarbon chemicals. The exact recipe is juggled to produce varying levels of fuel economy or performance. Lead was first added to gas in the 1920s to stop engines from "knocking" and as a lubricant, to help improve the durability of exhaust valves and seats. But lead in gas inevitably meant lead in exhaust emissions, and that meant lead – lots of it – in the environment.

Lead is highly toxic and, unfortunately, does not break down naturally in the environment. The lead poured out from motor exhausts each year simply adds to the lead already in the atmosphere. However, gasoline in Canada is being steadily purged of lead and the effects are dramatic. In the decade after 1974, when unleaded gasoline became widely available in this country, automotive lead emissions dropped from 12,980 tonnes to 7,278 tonnes.

Environment Canada has taken steps to get rid of the rest. In 1987, the Clean Air Act lowered permissible lead content in leaded gasoline to 0.29 g/L. Dovetailing with this act were regulations limiting emissions of hydrocarbons, carbon monoxide, and certain oxides of nitrogen, which required the use of catalytic converters (see below) and, by extension, unleaded gas in all new light-duty vehicles (cars with catalytic converters run only on unleaded gas). And as of December 1, 1990, leaded gas was no longer sold in Canada.

For detailed and up-to-date information on fuel efficiency, get a copy of the annual Fuel Consumption Guide: Ratings for new cars, pick-up trucks and vans, published by Transport Canada and available from most new car dealers, motor vehicle licence offices or by writing the Public Affairs Branch, Transport Canada, Ottawa, ON K1A 0N5.

* *

Catalytic Converters

T wo-way catalytic converters (presently the most common) help eliminate up to 90% of exhaust pollutants by converting hydrocarbons and carbon monoxide into water vapour and carbon dioxide. Three-way converters, which also oxidize certain nitrogen oxides, became mandatory with the 1988 models. Contrary to myth, they do not affect either your car's fuel economy or its engine performance.

Pollution abatement devices in use to date, however, tend to come with some strings attached, in this case the slightly increased levels of carbon dioxide and sulphur dioxide they produce. But it's a worthwhile trade-off, at least until better fuels – renewable and non-polluting – come along.

Similarly, lean-burn engines, which have higher-than-average air-to-fuel ratios, burn fuel more efficiently and reduce nitrogen oxide and carbon monoxide emissions at the expense of tending to increase hydrocarbons.

Each year the average Canadian car spews out:

- **60 kg of nitrogen oxides**

- **35 kg of hydrocarbons**

- **4064 kg of carbon dioxide**

Methanol is usually manufactured from natural gas, wood, coal, oil, or organic wastes.

Ethanol is distilled from various renewable feedstocks including corn, grain, and wood.

The Diesel Deal

Diesel cars are good examples of environmental trade-offs. On the one hand, in urban conditions they're much more energy-efficient than gas-powered vehicles, they're sturdier, they normally emit much less carbon monoxide and hydrocarbons, and they're lead-free. On the other hand, they produce more particulate pollution (soot or dark smoke), nitrogen oxides, and sulphur dioxides than gas engines with pollution devices, and they're noisier.

Overall, however, diesels still come out well – as long as you don't do a lot of highway driving. And if you opt for a diesel, be sure to have it regularly serviced.

✳✳✳✳✳✳✳✳✳✳✳✳✳✳✳✳✳✳✳✳✳✳✳✳✳✳✳✳✳✳✳✳✳✳✳✳✳✳

Alternative Fuels

Eventually we'll have to replace the gas in our cars with a non-polluting, renewable energy source. In the meantime, a number of alternate fuels with some advantages over gas are being tried, at least experimentally, with varying degrees of success. But so far all have drawbacks that chip away at the gains. It's a tricky issue.

The two alternatives most widely available in Canada are natural gas and propane. However, both are derived from non-renewable fossil-fuel sources. These high-octane fuels burn more cleanly than regular gas (they cut carbon monoxide and hydrocarbon emissions by 90% and are lead-free), are 6% to 15% more efficient, and are easier on engines. They may be an option for you if you do a lot of driving and live in an urban area where supplies and prices are good. The cost of converting a car ($1,500 for propane, $2,300 for natural gas) is often soon recovered in fuel savings.

Dedicated engine systems (those that run only on propane or natural gas) are somewhat cleaner than dual-fuel systems (those that also run on regular gas or blends): they reduce carbon dioxide emissions, whereas a dual system may actually increase them. However, in the case of natural gas, the carbon dioxide reduction may be partly offset by higher discharges of methane, a potent greenhouse gas.

A number of "biomass" fuels – those derived from plants – have been tried, most of them alcohols such as methanol ("wood alcohol") and ethanol (the kind in alcoholic drinks). In this country, methanol is usually manufactured from natural gas (Canada has a plentiful supply), but it can also be made from wood, coal, oil, or organic wastes. On the plus side it's a high-octane fuel that cuts noxious fumes in half and to some extent reduces carbon dioxide emissions (unless derived from coal, which would double them); and it can be blended up to 5% with gas without requiring changes to a car's fuel system or engine. On the minus side, it's currently very expensive to produce and is unlikely to be made in significant amounts from renewable sources.

Ethanol is distilled from various renewable feedstocks including corn, grain, and wood. In Canada it's selling mostly in the west in various "gasohol" blends that are much cleaner than gas. Carbon monoxide is cut by up to 30%. Unfortunately, ethanol production is expensive at present, although the technology exists to make it much more cheaply. As foreign markets for Canadian grain deteriorate, the ethanol industry may offer a new market for Canadian farmers.

Electric-powered vehicles, which are quiet and release no pollutants, are being tested experimentally. Here, the net environmental effects depend on what's used to generate the electricity. If it's coal, for instance, cars would release substantial amounts of carbon dioxide. Electricity is also likely to be expensive.

Experiments with solar power have so far yielded cars that are as light as bicycles and go about as fast. However, the main use of solar modules is not to drive the vehicle directly but to use solar electricity to recharge the battery of the electric vehicle.

Hydrogen, which you get by separating water, via electricity, into hydrogen and oxygen, may be our best hope for the (distant) future. It releases no pollutants, and unless the source of the electricity is fossil fuels, no carbon dioxide – its only emission is steam. It's also 15% to 45% more energy-efficient than gas. The catch, again, is the source of the electricity. If it's photovoltaics, wind, hydropower, or geothermal power, the environmental cost would be minimal.

* *

Cars of the Future

Improved car design has already made lighter, sleeker, more efficient models available to Green Consumers. But the fuel consumption of even the latest and best cars could be halved with the use of such innovations as more light materials like fibre-reinforced plastics and aluminum in bodywork, more ceramic components in the engine, and microprocessor-based fuel management systems. The fuel saved by using lightweight materials would probably not be offset by increased energy consumption in manufacture.

Several major car manufacturers (such as Volkswagen, Peugeot, Toyota, and Volvo) already have prototypes ready for production that can reach fuel ratings as low as 2.3 to 3.5 L/100 km. The problem is, they're still just prototypes. It seems that lower oil prices are encouraging car companies to slow down the incorporation of advanced fuel-economy technologies into mass-produced cars.

* *

The Environment-Friendly Driver

The car – whether gas-wasting behemoth or the most squeaky-clean model on the market – is only half the environmental equation. The other half is the driver. If you want to reduce to a minimum your car's negative impact on the environment, how you drive can be as important as what you drive. These driving and handling tips can help save not only the planet but your car and perhaps your life.

▬ Slow down. It's the simplest way to save energy and reduce pollution. An aggressive driving style characterized by high speeds and lots of sudden stops and accelerations drastically increases exhaust emissions and fuel use. Cutting your driving speed from 112 km/h to 80 km/h reduces fuel consumption by 30%; reducing speed also cuts nitrogen oxide emissions.

▬ Keep your car properly tuned and serviced to ensure maximum efficiency and minimum pollution. Despite the fact that manufacturers have designed better cars to comply with government emission standards, studies show that most Canadian cars still fail to meet them simply because they are not being adequately maintained.

▬ Fit radial tires and keep them inflated. By cutting tire drag, radials give you a 6% to 8% fuel saving. Under-inflated tires (check your owner's manual for the recommended pressure) wear out faster and can cost you 4% more for fuel.

◼◼◼ Avoid air conditioning, which uses CFCs and increases fuel consumption by 8% to 12%, especially in stop-and-go city traffic. Use the air vents and windows instead. A folding cardboard shade for the windshield will help keep the car cooler when it's parked in the sun.

◼◼◼ Consider car pooling as a cheaper, cleaner, energy-saving way to get to work. If you only need a car occasionally, rent one or take a taxi.

◼◼◼ Cover several errands in one car trip by doing some advance planning.

◼◼◼ Drive less. Most car trips are within five or ten kilometres of the driver's home. Whenever possible walk, bike, or take public transit.

* *

GREEN CONSUMER RECOMMENDED BUYS

The following cars have the best fuel-efficiency ratings in Canada, according to the *1990 Fuel Consumption Guide*. Copies of this handy guide are available from the Public Affairs Branch of Transport Canada (Ottawa K1A 0N5), as well as most motor vehicle licence offices and participating new car dealers.

Model	Transmission	City L/100 km	Highway L/100 km
Chevrolet Sprint	M5 +	5.5	4.4
Chevrolet Sprint	A3	5.9	5.3
Chevrolet Sprint Turbo	M5 +	6.3	5.0
Dodge Colt 100	M4	7.1	5.6
Eagle Vista	M4	7.1	5.6
Ford Festiva	M5 +	6.6	5.1
Honda Civic	M4 +	7.0	5.8
Mercury Festiva	M5 +	6.6	5.1
Nissan Micra	M5	6.6	4.8
Plymouth Colt 100	M4	7.1	5.6
Pontiac Firefly	M5 +	5.5	4.4
Pontiac Firefly	A3	6.0	5.4
Pontiac Firefly Turbo	M5 +	6.3	5.0
Subaru Justy	CV	7.0	6.0
Suzuki Swift	M5 +	6.2	4.9
Suzuki Swift Sedan	M5 +	6.4	4.9
Volkswagen Golf Diesel	M5 +	6.5	5.0
Volkswagen Jetta Diesel	M5 +	6.5	5.0
Volkswagen Jetta Turbo Diesel	M5 +	6.4	5.0

Codes: A3 — Automatic, 3 gears;
CV — Electronic Constant Variable Transmission;
M4 — Manual, 4 gears;
M4 + — Manual, 4 gears plus overdrive;
M5 + — Manual, 5 gears plus overdrive

An aggressive driving style can drastically increase exhaust emissions and fuel use.

Urban Planning for a Greener Future

The automobile has transformed a lot more than our atmosphere; it has changed the face of our cities and suburbs, and the way we live in them.

Modern subdivisions, usually built on former farmlands, would be unlivable without the cars that their residents use to escape them. Subdivisions have few large parks, no corner stores, no neighbourhood pubs or restaurants. Instead they have malls, and recreation areas that the kids have to be driven to. Their lack of sidewalks gives the game away: no one is expected to walk in these non-neighbourhoods.

Forward-looking planners can design new communities to reduce automobile use by adding sidewalks, local gathering places of all kinds, and public transit routes.

In the process they'll be encouraging the neighbourly interactions that make residential areas safe and attractive, as well as reducing traffic jams, car accidents', smog, and the acceleration of global warming.

PUBLIC TRANSPORTATION

Ranking right up there among the country's major polluters are those who live – and drive – in the fast lane: Canada's urban motorists. They consume over 40% of the transportation sector's petroleum, making a sizeable contribution to acid rain and global warming.

More than 75% of Canadian municipalities have a public transportation system, which doesn't leave city drivers with much of an excuse. If just one out of ten switched to public transit, annual global oil production could be cut by 17%. Mass public transit consumes far less energy than the automobile. According to Transport 2000 Canada, this is how the figures stack up:

Vehicle	BTUs/passenger-km
Subway or light rail (200 passengers)	175
Bus (67 passengers)	285
Small car (1 passenger)	2,570

Most drivers rolling along city streets are en route to work or the hardware store; in other words, taking short trips around the corner or across town. The assumption is that the car is faster and more convenient than a bus or streetcar. In Canada's big cities, that's not true any more. Traffic jams and congestion have slowed urban drivers down to a stop-and-go gridlock on most major streets and roadways. That means more stress on the gas

Traffic congestion means more stress on the gas tank, on the environment, and certainly on the human spirit.

* * * * * * * * * * * * * *

In Vancouver, the Bicycle Advisory Committee realized that real change requires a city bicycle plan, implemented by a full-time staff. Vancouver's Comprehensive Bicycle Plan, which has been adopted by the city council, is based on the "4Es":

- *Encourage* cycling.
- *Educate* cyclists and motorists.
- *Engineer* roads and other structures to facilitate safe cycling.
- *Enforce* laws that apply to cycling.

* * * * * * * * * * * * * * * * *

Bicycling in Traffic

Many Canadians are frankly scared about riding on crowded downtown streets. However, cyclists that have the proper skills and training ride safely in heavy traffic every day — and even say it's fun. To answer this training need, the Canadian Cycling Association has developed the CAN-BIKES SKILL program to teach the skills needed to avoid accidents and ride safely and comfortably in the city. Both introductory and advanced traffic cycling techniques are taught. For more information, contact the Canadian Cycling Association (1600 James Naismith Drive, Gloucester, ON K1B 5N4; (613) 748-5629).

Another good guide to traffic safety and important cycling principles is John Forester's book, Effective Cycling (MIT Press, 1984).

* * * * * * * * * * * * * * * *

Buy a Bike, Then Buy a Helmet!

Studies in Canada, Australia, and the U.S. show that cycling mishaps are the leading cause of hospital admissions for head injuries among school-age children. Yet only 2% to 5% of riders wear helmets. If you ride a bike, buy a snug-fitting bicycle helmet approved by the Canadian Standards Association (CSA) — and wear it.

* * * * * * * * * * * * * * * *

tank, on the environment, and certainly on the human spirit.

Public transportation has other advantages. Per passenger, a bus takes up about one-ninth of the road space occupied by cars, which means less congestion and less pollution. As well, building and maintaining city thoroughfares requires enormous quantities of land and resources, two more environmental problems that could be greatly reduced if more drivers were to abandon the car for short, in-city trips. The expansion and maintenance of a public transit system also requires land and resources, but it's certainly one of the better trade-offs.

In some of the country's urban centres, public transportation systems are just as congested as the city streets. For instance, in Toronto and Montreal the systems are already overtaxed, and city buses in Edmonton and Calgary are reaching a saturation point. It's hard to attract more riders to these overcrowded and often unreliable transit systems. What's needed is some lobbying for improvements on the part of riders and some vision and foresight on the part of city planners and politicians. To encourage the switch from private to public modes of transportation, the goal should be an efficient and integrated transportation system that offers commuters as many alternatives as possible.

For downtowners, walking and public transit are the best choices. For people who live in the suburbs, it may be a little more complicated. Ride-sharing and driving to the closest transit centre are possible alternatives. And one option that needs more serious consideration by both commuters and planners is the bicycle.

* *

The Bicycle

With the automobile destroying the environment and the quality of urban life, it's time to reconsider the advantages of the bicycle. Topping the list is its energy efficiency, which far outshines anything else on the road. A 25-km round trip on a bicycle burns 350 calories of energy, about as much as a person consumes in a hearty breakfast. The same trip by car requires 18,600 calories of energy (and in a less renewable form than bacon and eggs).

Another fact: in a car, one litre of gasoline provides 10.5 passenger-kilometres; in a city bus, about 42 passenger-kilometres; on a bicycle, the energy equivalent of more than 425 passenger-kilometres.

North Americans unfortunately consider the bicycle child's play. Obviously, we have lessons to learn from elsewhere. In Asia, the bicycle is the major mode of in-city transportation, and a few European countries launched national campaigns in the 1980s to encourage commuters to make the conversion from four wheels to two. In the Netherlands, for instance, there are almost 20,000 km of well-used bicycle paths; about half of all trips in the country are now made on two wheels. When the town of Erlangen in West Germany created 250 km of two-wheel paths, bicycle commuting doubled.

The same kind of effort and commitment has to be made by planners and politicians in Canadian cities. Right now, bicycle commuting is simply too dangerous on congested city streets. Besides bicycle paths, we also need bike-and-ride centres to encourage suburbanites to make the switch. What would also help are two-wheeler parking areas, as well as shower facilities. Once again, the bottom line is an integrated transportation network that provides space and facilities for the cyclist.

* *

Intercity Travel

H ow we travel from Moncton to Moose Jaw is as environmentally important as how we travel from home to work. Once again, energy efficiency per passenger-kilometre is the key issue:

Vehicle	BTUs/passenger-km
Train	270
Bus	320
Small car (2 passengers)	1,000
Boeing 767	1,990

It's clear from this chart that buses and trains are the most efficient modes of cross-country travel. Unfortunately, we've been abandoning them in droves in the last few years. Here are some alarming statistics from Transport Canada.

■ From 1984 to 1987, the number of passengers choosing air travel steadily increased, reaching 18.9 million in 1987.

■ In 1981, 8 million Canadians travelled by Via Rail; in 1987, the number of passengers had fallen to 5.9 million, a decline of 26.3%.

■ The total number of intercity public transportation passengers declined from almost 60 million in 1980 to about 46 million in 1987, with buses losing about 90% of the shortfall.

Cheaper fares and better service from the airlines, as well as more money in the pockets of passengers, account for part of the decline in train and bus travel.

But something else is afoot here. Most Canadians find travelling more than about 500 km by bus too cramped and uncomfortable; their mode of choice is Via Rail, which stops in 900 Canadian cities, towns, and isolated communities (only 200 of these have airports). But with Via's service steadily deteriorating over the last decade, ridership has dropped. A large part of the problem is the federal government, which has abandoned Via for more energy-intensive modes of public transportation, especially air and road travel.

According to Transport 2000 Canada, studies indicate that 85% of Canadians believe that improving Via Rail is a "federal obligation." What it will take is some intense and relentless public lobbying for improved service and modern equipment. For the average consumer, a good start is to support Transport 2000's ongoing campaign to get the country back on the rails (**Transport 2000 Canada**, 22 Metcalfe Street, Suite 405, Ottawa ON K1P 5P9).

The motor oil that your car can't use any more can be re-refined for reuse. That saves a non-renewable resource, as well as sparing the environment the effects of improper disposal. Used oil should never be burned, poured into a sewer, or dumped into landfill: along with oil go toxic and carcinogenic chemicals such as benzene, lead, cadmium, and polynuclear aromatic hydrocarbons.

If you have your car's oil changed at a service station, ask the manager what happens to it. Many stations have their oil collected by a contractor; will the contractor dump it or refine it?

Only a few Canadian communities have public collection systems for do-it-yourself oil-changers who want to have their used oil re-refined. Ottawa residents can pour their oil into a clean, sealed plastic container and put it out beside their Blue Box. In Metropolitan Toronto and in Alberta, certain service stations collect the oil.

If you do your oil change yourself and you don't live in an area offering a used oil collection or drop-off service, it will be harder to get the used oil into the hands of re-refiners. Most contractors who do this for service stations will not deal with individual consumers (check your Yellow Pages for "waste oil" or "oils-used" for more information).

But at least you can ensure that your oil is disposed of properly. Put it with your other household hazardous waste and treat it accordingly (see Chapter 7).

And when you need more oil, look for a re-refined motor oil.

9. WORKING and INVESTING

S. Quinlan

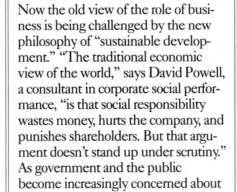

FOR MANY YEARS IT'S BEEN SAID *that the only business of business is to make money. Other sectors could worry about social or ethical issues; the best way for business to benefit society was to be successful. But lately, business has begun to realize that the companies that do good also tend to do well. Although they may not have perfect records, many of the most successful corporations in the world are also known for having good employee relations, supporting the arts, or sponsoring community projects.*

▲▲▲▲▲▲▲▲▲▲▲▲▲▲▲▲▲

Now the old view of the role of business is being challenged by the new philosophy of "sustainable development." "The traditional economic view of the world," says David Powell, a consultant in corporate social performance, "is that social responsibility wastes money, hurts the company, and punishes shareholders. But that argument doesn't stand up under scrutiny." As government and the public become increasingly concerned about pollution, and the fines begin to

mount, "it's a sign of stupidity if a business refuses to take environmental issues seriously," says Powell.

There are substantial initial costs in being an environmentally responsible company – costs that are not incurred by less responsible corporations, especially those operating in countries where pollution laws are lax. Canadian industry feels that companies that take costly environmental measures should be compensated by tax breaks. That suggestion is controversial: should the public pay for what companies should be doing anyway? But we've all benefitted from the success of these companies, counters business, and bought their products. And if prices rise to pay for pollution controls, Canadian business will lose out to international competitors. It's not an easy issue to solve.

❝ Canadian industry feels that companies that take costly environmental measures should be compensated by tax breaks. Should the public pay for what companies should be doing anyway? **❞**

F ortunately, some companies have found that investments in pollution control have paid off generously. One of the first companies to discover this was Minneapolis-based **3M**. In 1975 3M instituted its "3P" program – "Pollution Prevention Pays." It did. Through recycling, the early introduction of pollution-control equipment, and eliminating pollution from the production process, the company saved $235 million over the next 11 years.

Even though pollution-control devices are a heavy investment, many companies are finding ways to allay the cost. Through the **Canadian Waste Materials Exchange** in Mississauga, Ont., hundreds of companies are saving money by swapping waste chemicals, solvents, and other waste materials that would otherwise become pollutants. **Inco Ltd.** designed its own system to cut sulphur dioxide emissions and is now selling the design to other firms. In Owen Sound, Ont., **RBW Graphics**, a printing company, decided to take a hard look at its waste management after the city announced that landfill rates would double. After about a year of recycling waste paper, ink, and other garbage, the company found it had saved $280,000 at a cost of well under $100,000.

S o doing the right thing environmentally doesn't necessarily mean losing money. In the past, both business and environmentalists tended to believe that the only way to preserve the environment was to limit growth. Now the concept of sustainable development, as outlined by the Brundtland Report, has suggested that, if we work at it hard enough, economic success and good environmental practices can go together. In fact, environmental needs can even stimulate business. For example, a large part of the international success of Japanese automakers stems from the more fuel-efficient engines they were obliged to develop in order to meet tough government emission standards at home. (American automakers responded to a similar situation by fighting the pollution controls in court.)

In fact, there are indications that preserving and cleaning up the environment will become a big money-making industry. Experts have predicted a growth rate of as high as 20% to 30%. One leader in the rapidly developing field of environmental problem-solving is **Eco-Tech Inc.** This wholly owned Canadian company designs and manufactures systems that recover chemicals and minimize hazardous waste in the metal-finishing industry. From a small company with only a couple of full-time employees, Eco-Tech has grown to become a leading supplier in its area, with 75 employees in Canada and a plant in Britain, exporting to 28 countries. Meanwhile, among new consumer products, "green" products have outperformed regular items, and industry experts estimate the North American market for environment-friendly products could be worth over $1 billion by the early 1990s.

There are indications that preserving and cleaning up the environment will become a big money-making industry. Experts have predicted a growth rate of as high as 20% to 30%.

GREEN WIDGET CO.

Employees can lobby their companies to set up environmental committees to monitor the company's performance.

But where does this leave the individual? We tend to think of corporations as huge faceless organizations that we cannot possibly influence. We forget that those companies are made up of employees and rely on investment by individuals to operate. In short, those companies cannot exist without us. So by influencing the decisions made by the companies we'll invest in, we can make real improvements in the way Canadian corporations treat the environment.

* * * * * * * * * * * * * * * * * * * *

Influencing Decisions

If you're the general manager of a petrochemical refinery, then of course you probably have a regular opportunity (and obligation) to make good environmental decisions. But even less exalted employees can take responsibility for their company's environmental policies.

Employees can lobby their companies to set up environmental committees to monitor the company's performance. Employees can also implement purchasing policies that consider the environmental records of the companies they purchase goods and services from. You may also want to encourage your company pension plan not to invest in companies that harm the environment. And because workers are often the first to feel the effects of chemicals, they can be effective in alerting employers and the public to environmental problems and public health.

There are also steps you can take in any office or factory, regardless of what your company actually does, to make your workplace more environment-friendly. If your premises are ripe for renovation, you will have even greater opportunities to implement energy-efficient improvements in such areas as heating, air conditioning, and ventilation; see Chapter 5.

* * * * * * * * * * * * * * * * * * *

10 Steps to Developing a Green Business Strategy

Many firms have learned that taking care of the environment makes good business sense. The Canadian Chamber of Commerce has formulated a 10-step plan that a company can follow to reduce its impact on the environment, improve its public image, and tap into green market opportunities. The following provides a thumbnail sketch of the program. For more information, obtain a copy of *Achieving Environmental Excellence: A Handbook for Canadian Business*, for $10.00 from The Canadian Chamber of Commerce (55 Metcalfe St., Ottawa, ON K1P 6N4; (613) 238-4000).

1. Document your corporate environmental goals in a one-page policy statement.

2. Appoint a senior company officer as your "Environment Champion" and set up a support team representing every major department in the firm.

3. Conduct an environmental performance review or audit of your company's operations.

4. Prepare a plan of action setting out specific tasks, deadlines, costs, and staff responsibilities.

5. Get your staff involved in making the plan a success.

6. Allocate sufficient funds and other resources to get the job done.

7. Conduct ongoing research — subscribe to environment publications, attend conferences and trade shows, join environment-related associations.

8. Communicate your environmental activities to your employees, your customers, and the public.

9. Adopt a spirit of cooperation among environmental interest groups.

10. Take a long-term view — some investments may take years to produce a financial return. Others may pay off in public good will.

✳✳✳✳✳✳✳✳✳✳✳✳✳✳✳✳✳✳

Recycling

Paper, cardboard boxes, plastics, cans, glass bottles, and food waste can all be recycled. One of Toronto's large downtown office complexes, the Sun Life buildings, recycled 180 of 780 tonnes of waste in 1988, reducing their disposal costs by $25,000. There are developing markets for a number of other common industrial, commercial, and industrial wastes. These include:

• Empty steel-fibre and plastic drums

• Drywall (which has been banned from some landfills)

• Asphalt, concrete, and clean fill

• Flat glass (in large enough quantities)

• Rendering wastes

• Laser printer cartridges

For more information on recycling wastes, contact the Canadian Waste Materials Exchange (c/o ORTECH International, 2394 Speakman Dr., Mississauga, ON L5K 1B3; (416) 822-4111).

✳✳✳✳✳✳✳✳✳✳✳✳✳✳✳✳✳✳

The Average Office Worker Wastes 73 Kilograms of Fine Paper Each Year

Here are five ways to save paper in the office:

1. Photocopy on both sides of the sheet.

2. Use waste paper for rough notes, messages, etc.

3. Replace your fax front sheet with a half sheet.

4. Make one copy of a memo and attach a routing slip.

5. Make more use of "phone-mail" and computer message systems.

✳✳✳✳✳✳✳✳✳✳✳✳✳✳✳✳✳✳

Buy Environment-Friendly Office Products

• Waste baskets and recycling bins made from recycled plastic

• Non-toxic "white-out" fluid and "high-lite" pens

• Recycled file folders

• Recycled white and commercial envelopes and padded mailing bags

• Recycled computer paper and photocopy paper

• Recycled letterhead and fine paper

• Recycled adding machine rolls

• Notebooks and legal pads, telephone message pads

• Recycled pressboard binders

Paper, cardboard boxes, plastics, cans, glass bottles, and food waste can all be recycled.

Paper Recycling

Paper recycling makes environmental sense. According to the Wisconsin Consumer Packaging Council, manufacturing recycled paper produces 74% less air pollution and 35% less water pollution than making paper from virgin wood pulp. In addition, recycling uses 58% less water and 64% less energy. And you needn't sacrifice quality to achieve these environmental benefits. Recycled papers are often more opaque, denser, and more flexible than virgin paper.

So naturally you'll want to purchase recycled paper. The paper you are dropping into the office recycling bin — the industry calls this "post-consumer" waste paper — isn't yet being recycled into new office paper, at least not in Canada. It's made into tissue and boxboard. But increasing demand is encouraging papermakers to invest in the equipment that will make it possible for Canadians to close the recycling loop.

In the meantime, lots of office paper with "recycled content" is on the market. Most manufacturers use the term to include a combination of two main categories of fibres. The first category is the paper and fibre that is left over during the regular milling process, which has always been reclaimed and recycled. These materials make up at least 10% of the content of most "new" paper. The second principal source of recycled content is "post-commercial" waste paper: printed paper from printing plants and similar pre-consumer sources.

****

Corrugated Cardboard

Many municipalities are now recycling corrugated cardboard; some have even banned it from their landfill sites. If your community recycles corrugated cardboard:

• Remove all plastic wrappings, metal and plastic straps, and wood or foam liners from the boxes;

• Flatten boxes;

• Bundle or stack them together; and

• Place the bundles curbside on your regular garbage pickup day, or contact the Public Works department for pickup details.

Lots of office paper and fine paper with "recycled content" is available in the market.

****

What Can We Do?

Waste	Percentage	3R Options	Recycled Products
Mixed paper (magazines, junk mail, phone books, grocery bags, packaging)	20%	Reuse bags, get off junk mailing lists, share magazines	Recycled paper products and construction materials (roofing shingles, tar paper, insulation, etc.)
Newspaper	15%	Share your paper, use the library, use TV/radio	De-inked newsprint, cellulose insulation, packing and building materials
Corrugated cardboard	3%	Reuse boxes	Cereal boxes, writing pad bases, file folders, wallboard, etc.
Fine paper/office paper	2%	Use both sides of paper, make only as many copies as you need, use routing slips, use electronic mail, set up an office fine paper recycling program	Writing and computer paper, paper towels, napkins, tissues
Glass bottles and jars	9%	Buy larger sizes, wash and reuse bottles, pick refillables, take your own container and buy in bulk	New glass products, glass asphalt and road base aggregate, fiberglass
Metals (aluminum cans and foil, tin/steel cans)	5%	Pick refillable or reusable containers (instead of pop cans), reuse foil, buy fresh produce, meat and fish instead of canned	New metal products, cans, window frames and construction products
Plastic (all types)	5%		
High density polyethylene (HDPE) used in milk and water jugs, detergent bottles, etc.		Use refillable glass bottles, wash and reuse plastic tubs, make your own cleaners or buy refills	Drainage pipes, flowerpots, toys, traffic cones, etc.
Polyethylene terephthelate (PET) used in soft-drink bottles		Pick refillables, don't drink pop	Carpet fibres, fill for pillows, sleeping bags, rigid urethane foams, engineered plastics such as floor tiles, shower stalls, drainage pipes, etc.
Polystyrene foam used in fast-food containers		Buy bulk, bring your own container, choose alternative packaging	Insulation, packing material, "plastic wood"
Yard wastes	25%	Use alternative ground covers instead of grass, leave grass clippings on the lawn, use home composter or support central composting programs	Compost mulch for lawns and gardens
Food scraps	10%	Plan meals, buy fresh food, make better use of leftovers, use home composter or support central composting programs	Compost
Disposable diapers	2%	Cloth diapers	Pilot recycling projects underway

(some of the above information was taken from the EHMI Recycling Wheel, P.O. Box 932, Durham, NH 03824)

For More Information on Greening the Workplace

- ☐ *Profit from Pollution Prevention — A Guide to Waste Reduction and Recycling in Canada*, 2nd Edition, for $25.00 from Pollution Probe (12 Madison Ave., Toronto, ON M5R 2S1)
- ☐ *Working for the Environment: Ideas for Workplace Improvements*, and
- ☐ *Environmental Considerations for Conferences and Meetings*, $2.00 each from the Harmony Foundation (P.O. Box 4016, Station C, Ottawa, ON K1Y 4P2)
- ☐ *Your Office Paper Recycling Guide*, Toronto Recycling Action Committee, and
- ☐ *The Office Guide to Waste Reduction and Recycling*, both available from the Recycling Council of Ontario (489 College St., Suite 504, Toronto, ON M6G 1A5)
- ☐ *Green Pages — Green Business Suppliers' Guide*, MAB Designs (741 Garyray Drive, Weston, ON M9L 1R2)

Lighting and Energy

Use compact fluorescents in both overhead and desk lamps, and make sure there are accessible switches so they can be turned off when they're not needed. Halogen lights can be used as a brighter source of desk or task lighting, allowing overall light levels to be lowered. Reduce energy consumption by turning off lights and idle machines.

Cleaning

Substitute less toxic cleaners for the usual commercial products (see Chapter 3). See that the old cleaners are treated as hazardous waste when they're disposed of.

* * * * * * * * * * * * * * * *

Washrooms

Make sure you have aerated faucets (see Chapter 5) and reusable cloth roll towels. Tank dams in toilets will save water.

* * * * * * * * * * * * * * * *

Kitchens and Cafeterias

Use washable dishes and cutlery instead of disposables. Even if you've only got a coffeemaker, you can do your part by using mugs and spoons instead of disposable cups and stir sticks. Make sure your refrigerator is efficient, and generally follow the same rules as for your kitchen at home (see Chapter 5).

Office Air Quality

Even low levels of ozone gas can be a serious respiratory hazard as they build up in a stuffy office. Emissions from photocopiers and laser printers can easily exceed ozone levels of 1.5 parts per million — a level that reduces vital lung capacity by 20%. Adequate ventilation can help reduce indoor air pollution levels.

Landscaping

If you're lucky enough to have lawns and flowerbeds around your workplace, discourage the use of chemical fertilizers and pesticides. Food waste from the cafeteria can be composted to improve the garden's soil.

Transportation

Provide incentives for employees to carpool or take public transport to work. If your company supplies cars to employees, encourage them to purchase fuel-efficient vehicles with good pollution controls. Equip company vehicles with retreaded tires and buy re-refined motor oil. Inspect and maintain emission control devices in company vehicles. To encourage cycling, install bike racks and provide showers. Hire a bicycle courier where possible.

* *

WHMIS

The Workplace Hazardous Materials Information System (WHMIS) was designed to ensure that workers know how to handle hazardous materials safely. It applies to all workplaces in which such materials are used. The basic elements of a WHMIS program are strict regulations for labelling every hazardous material used in a plant; data sheets for every hazardous material, giving further information on how to use it safely; and education for *every* worker in the plant.

WHMIS requires that all workers be educated on the effects and dangers of the hazardous materials they may be exposed to on the job, proper storage and disposal protocols, emergency procedures, and what special precautions they need to take, including the use and maintenance of personal protective equipment ent (gloves, safety glasses, respirators, etc.). A pictogram labelling system makes sure that even workers who are illiterate or who do not read English will be able to understand the dangers of the materials they're handling. Although WHMIS increases the chemical knowledge of individual workers, it does not absolve companies of their duty to make sure that workers are properly educated and that chemicals and other hazardous materials are being handled properly.

* *

Whistle-Blowing

There are times when employees are unable to change a company policy or action that they know is having a harmful effect on the environment. If you find yourself in that situation you may want to consider whistle-blowing – informing the public and/or the government of the company's misbehaviour.

Think carefully about that decision. All provinces have legislation protecting whistle-blowers – an environmental protection or health and safety act – *if* they take their information to the government. But your legal situation if you're caught and punished by the company is far from clear-cut, and it may prove expensive to defend yourself.

If you don't belong to a union and if there's any other good excuse for firing you – the company is overstaffed and you're the least senior person, for example – then you're particularly vulnerable. Moreover, people who inform on their company, even for the most righteous cause, are often ostracized by co-workers, friends, and relatives and may suffer serious financial and personal problems. The name Karen Silkwood is enough to remind us how serious the consquences can be for a whistle-blower. Nor are results guaranteed. As one whistle-blower said, "If you have God, the law, the press, and the facts on your side, you have a 50-50 chance of winning."

Whistle-blower protection varies from province to province and, in some cases, requires that employees report environmental or occupational health problems to provincial authorities or face charges themselves. For further information and advice on whistle-blowing, you can contact the Canadian Environmental Law Association (517 College St., Suite 400, Toronto, ON M4G 4A2) or the Canadian Labour Congress (2841 Riverside Drive, Ottawa, ON K1Z 8X7).

Some people feel that brown-enveloping and waiting for the right moment are cowardly. But the important thing is not for you to play the hero, it's to expose the danger.

Ethical Investing

As companies become more involved in the business of "saving the environment," the idea of individuals using "ethical investments" to support environmentally responsible concerns becomes very attractive. There are many different ways of investing your money, from leaving it in a bank account to buying a home. But usually "ethical investment" refers to buying stocks and bonds of publicly owned companies. Ethical investors refuse to invest in corporations whose policies they disagree with and may also try to encourage responsible corporations by buying their bonds and shares.

Of course, deciding which companies are environmentally responsible is not easy. Should you invest in a company that continues to pollute the environment but is making measurable changes to improve its environmental record? Or should you keep yourself pure and invest only in non-controversial companies – while continuing to drive your car and use unrecycled paper products? Do you simply want to avoid companies with a bad environmental record? Or do you want to actively encourage companies that are showing environmental initiative? Since environmental concern has become very fashionable, how do you tell the sincere corporation from the one that's cynically donned a few flashy improvements to cover a less-than-impressive performance? And do you really have the financial security, temperament, and interest to invest in corporations? Perhaps you'd be better off making your own ethical statement by investing in a community co-operative fund or a credit union.

An invaluable guide for dealing with all these issues is *The Canadian Guide to Profitable Ethical Investing* (Lorimer, 1989). The guide points out the ethical implications of every kind of investment, from savings accounts at your local bank to credit unions to stocks and bonds, as well as explaining how those investments work.

Some investors take a more proactive approach and buy stock in a company in order to have a say in how that company is run. Churches, universities, and other concerned organizations have often used their stock holdings to try to influence a corporation's policies. Certain types of stock entitle you to voting rights at a company's annual meeting. It's highly unlikely that you as an individual will ever have enough stock to vote down a bad policy, but your votes do give you the right to speak at the annual meeting.

If you've really got big bucks to play with, then you might want to consider lending "venture capital" – start-up funds – to companies with exciting environmental technology. You could be funding the ecological equivalent of the Gutenberg press, but you could also lose everything you've invested, so this idea is obviously only for people who are financially secure. **Envestment Corporation** is attempting to match investors with environmental companies that need venture capital. You can reach them at 47 Lynngrove Ave., Etobicoke, ON M8X 1M7; (416) 234-5227.

If you do want to invest in corporations, then you have to decide whose stocks and bonds you want to buy. In response to public interest, the financial community is becoming more sensitive to ethical issues. **EthicScan Canada Ltd.** (89 Edilcan Drive, Concord, ON L4K 3S5) is one company that will provide information on companies related to a variety of issues, including the environment, for a fee. The company also publishes a bi-monthly newsletter, *The Corporate Ethical Monitor.*

Another book that can help you decide what companies you want to invest in is *Rating America's Corporate Conscience*, written by the Council on Economic Priorities (Addison-Wesley, 1986). (A shorter guide from the Council, *Shopping for a Better World: A Quick and Easy Guide to Socially Responsible Supermarket Shopping*, contains similar information, but with much less detail.) Although the book deals only with American corporations, many of these companies have branches or subsidiaries in Canada. Be aware, however, that a company's environmental record may be different here than it is in the U.S. For further information you can contact: **Council on Economic Priorities**, 30 Irving Place, New York, NY 10003.

There are also a few environment companies listed on stock exchanges. But remember that many new technologies, such as biodegradable plastics, are controversial. Moreover, these stocks can be volatile, especially when some new environmental crisis is at the top of newscasts.

If you feel you don't have the time or expertise to decide what companies you should invest in, consider a mutual fund. When you buy shares in a mutual fund, the fund takes the money placed in it by investors and invests it in a variety of companies. Mutual funds have become popular recently because they are seen as a relatively safe way for investors to play the stock market, and they have a fairly good rate of return. And since someone else makes the decisions about what companies to invest in, the mutual fund saves the investor a lot of research time. (Remember, however, that unlike a savings bond or bank deposit, there is no insurance on a mutual fund. If the fund declines drastically in value or the fund's managers prove unscrupulous, you have no protection against losses.)

Mutual funds are seen as a relatively safe way for investors to play the stock market and have a fairly good rate of return.

* *

Getting into Mutual Funds

Some good ethical mutual funds have decided not to include environmental factors in their criteria for companies they will invest in, because they feel the issues are too complicated. Instead, they stick to issues such as a corporation's dealings with South Africa and Third World countries or its involvements in military contracts. (They may also decline to invest in companies involved with nuclear power.) However, the mutual funds listed below do screen companies for their environmental policies. Some funds simply avoid companies with bad environmental records, while others actively seek out corporations with innovative environmental programs. (You will probably need to make special arrangements through your broker to purchase the U.S. funds.)

GREEN CHIP STOCK

Ethical mutual funds in Canada still have only a tiny percentage of the market for mutual fund investments, but the movement is fast-growing and could eventually control 10% of the $150 billion in Canadian pension funds — not to mention individual investments. For more information, contact The Social Investment Organization (366 Adelaide St. E., Suite 447, Toronto, ON M5A 3X9), a non-profit group that promotes ethical and alternative investment in Canada.

✳ ✳

These mutual funds do screen companies for their environmental policies.

These mutual funds screen companies for their environmental policies:

▬ Environmental Investment Canadian Fund/International Environmental Investment Fund (Sponsored by Energy Probe), Environmental Investment Funds Ltd., 225 Brunswick Avenue, Toronto, ON M5S 2M6

▬ Investors Summa Fund Ltd., 280 Broadway, Winnipeg, MB R3C 3B6

▬ Dreyfus Third Century Fund, 767 Fifth Avenue, New York, NY 10153

▬ New Alternatives Fund, 295 Northern Boulevard, Great Neck, NY 11021

▬ Pax World Fund, 224 State Street, Portsmouth, NH 03801

▬ Ethical Growth Fund, VanCity Investment Services Ltd., 515 West 10th Avenue, Vancouver, BC V5Z 4A8

In addition, the following funds are open to group investors, such as pension funds:

▬ Canadian Ethical Dynamic and Responsible Balanced Fund (CEDAR), #1, 10005 – 80 Avenue, Edmonton, AB T6E 1T4

▬ Crown Commitment Fund, Crown Life Insurance, 120 Bloor St. E., Toronto, ON M4W 1B8

Theoretically, ethical investment funds should not perform as well as funds with no restrictions, because they have a narrower range of companies to invest in. Contrary to theory, however, most of the funds have performed quite well. But the performance of particular funds or of mutual funds in general may change over the years, so make sure you check a fund's recent performance before investing and track it carefully once you have invested. A broker or financial planner may be able to help you.

✳ ✳ ✳ ✳ ✳ ✳ ✳ ✳ ✳ ✳ ✳ ✳ ✳ ✳ ✳ ✳ ✳ ✳

For further information on ethical investing and the environment, you can contact the following organizations.

▬ The Social Investment Forum, 711 Atlantic Avenue, Boston, MA 02111

▬ Taskforce on the Churches and Corporate Responsibility, 129 St. Clair Avenue West, Toronto, ON M4V 1N5

▬ Social Investment Study Group, c/o Ted Jackson, E.T. Jackson and Associates, Suite 712, 151 Slater Street, Ottawa, ON K1P 5H3

I f you decide to get involved in ethical investing, remember that good intentions are no guarantee that you won't lose money. Investments you consider ethical are subject to the same fluctuations as any other investment. So do your research, get expert advice from a stockbroker or financial planner, and be aware of any risks you're taking.

* *

AFFINITY CREDIT CARDS

A nother way to send money in the right direction is through "affinity" credit cards. These are special VISA cards or MasterCards that direct a portion of your payments to a designated charitable organization. Every time you purchase something with the card and every time you pay an annual fee, the bank, trust company, or credit union donates a percentage of that money to the organization. For example, both Bank of Montreal and Canada Trust (through their non-profit Friends of the Environment Foundation) now offer affinity MasterCards that provide funds to environmental groups. Investors Group Trust Co. Ltd. has launched a new VISA credit card — the Environment Card. And in Vancouver, VanCity Savings Credit Union has set up an EnviroFund to support local environmental initiatives; when VISA card holders use their cards at least once a month, VanCity makes a donation to the fund. If your credit card funds an operation you want to support, why not switch your card to an affinity card? If it doesn't, then encourage your bank, trust company, or credit union to fund that organization — or switch your credit card to one that does.

Credit Unions are a $50-billion alternative financial system — they are the single largest resource for people who want to put their investment dollars to work in their own community. For more information, as well as a listing of credit union organizations in each province, contact the Canadian Co-operative Credit Society (300 The East Mall, 5th Floor, Islington, ON M9B 6B7; (416) 232-1262).

* * * * * * * * * * * * * * * * * *

DOES YOUR BANK…

- finance nuclear power?
- finance clear-cutting of forests?
- finance production of toxic chemicals?
- finance paving over of farmland for urban sprawl?
- support organic farming?
- have a fine-paper recycling program?

(Source: Bread & Roses Credit Union, 348 Danforth Ave., Suite 211, Toronto, ON M4K 1N8)

INVEST IN GREEN

10. TRAVEL and LEISURE

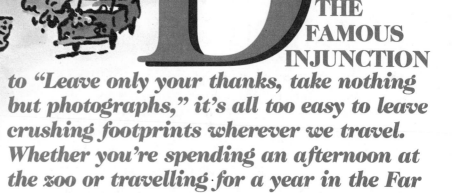

DESPITE THE FAMOUS INJUNCTION to *"Leave only your thanks, take nothing but photographs,"* it's all too easy to leave crushing footprints wherever we travel. Whether you're spending an afternoon at the zoo or travelling for a year in the Far East, there are environmental issues to be considered. Your choice of transportation is important; see Chapter 8. How will you affect the cultural and physical environment you're entering? And while you're having fun, are you supporting practices and organizations you'd rather not be involved with?

▲▲▲▲▲▲▲▲▲▲▲▲▲▲▲▲▲▲

"It appears that many people when they travel really see nothing at all except the reflection of their own ideas."

Stephen Leacock

▲▲▲▲▲▲▲▲▲▲▲▲▲▲▲▲▲▲

Hundreds of millions of people now travel internationally every year, and Canadians are eager members of this globe-trotting family. Exotic destinations are becoming increasingly popular. By 1984 about 17% of all international travellers were visiting the Third World — twice the proportion of 10 years before. Nor is it only young backpackers who are trekking in Nepal or cruising the Galapagos Islands. Third World countries are becoming standard destinations for more conventional travellers. This trend is causing problems for both the physical environment and the people who live in those countries.

> **At its best tourism can be an instrument of greater understanding, breaking down the barriers among people and informing them about one another's culture, society, and philosophy**

Contours, the newsletter of the **Ecumenical Coalition on Third World Tourism** (P.O. Box 24, Chorakhebua, Bangkok 10230, Thailand), outlines the problems this trend is creating:

The geographical direction of tourism, on an international level, consists of people from the First World travelling to visit islands, beaches and picturesque towns of the Third World. But though we opt for the "tropical paradise," we demand it come with all the comforts of home. So we drop glass-and-steel hotels next to peasant villages or fishermen's huts, creating luxuriant, forbidden facilities that no resident would dare enter… Though we may be only modestly well-off, we pay prices of a staggering size in local terms… In the course of our stays, we make no use at all of the lodgings that residents use, or of their transportation or dining spots.

Although tourism is often promoted as a way of bringing money into poorer nations, it may actually be economically disadvantageous for a Third World country. Consider that money has to be sent out of the country to pay for elevators, cars, buses, air conditioners, Western-style food, and the other amenities that tourists demand. Much of the money tourists spend is paid up front to foreign tour companies, airlines, and hotel chains before they even leave home. But meanwhile the government is spending huge sums on tourism development in the hopes it will bring in foreign currency – sums that might better be spent on health care and education or other forms of development. Land must be cleared – often without proper recompense and sometimes with violence – for hotels, airports, and tourist sites. Farmers and fishermen give up their traditional livelihood and become dependent on the tourist trade; but the best jobs usually go not to local people but to staff imported by the international companies. Traditional social structures and ways of life vanish, resulting in social problems. And if there's a hurricane, political unrest, or simply a change in fashion, the tourists stop coming and suddenly the community is bereft of the business it's come to rely on, with little to fall back on.

The incidence of economic dislocation, prostitution, and other havoc wreaked by tourism in the Third World is enough to make you feel you should just stay home. But at its best, tourism can be an instrument of greater understanding, breaking down the barriers among people and informing them about one another's culture, society, and philosophy.

One way to make travel a method of international diplomacy rather than economic exploitation is to live more like the locals. Instead of staying in the big international hotels and eating in expensive Western-style restaurants, stay in local lodgings and experiment with the local food. That way, you're supporting the local economy.

Make travel a method of international diplomacy: stay in local lodgings, experiment with local food – support the local economy.

Travelling closer to the ground is not only cheaper, it's more stimulating. You're exposed daily to a different culture, a different way of looking at the world. No less a person than Arthur Frommer of *On $5 a Day* fame has come to the same conclusion. "After 30 years of writing standard guidebooks," writes Frommer in the introduction to his latest guidebook, "I began to see that most of the vacation journeys undertaken by Americans were trivial and bland, devoid of important content, cheaply commercial, and unworthy of our better instincts and ideals."

In his *The New World of Travel 1989* (Prentice-Hall, New York, NY), Frommer takes a more authentic approach to travel, covering everything from ethical, pro-environment travel and study vacations to discount travel and tours. Other travel books, such as the *Lonely Planet* series, give good advice on travelling cheap in the Third World. You can get *Alternative Tourism: A Resource Guide* and other publications on responsible travelling from the Center for Responsible Tourism, 2 Kensington Road, San Anselmo, CA 94960, USA.

Let's not be naïve, however. Travellers who stay and eat in local establishments are just as capable of bad behaviour as those staying in air-conditioned international hotels – and the local people are forced to put up with *both* groups. So if you're going to travel that way, be on your best behaviour. Improper dress, stinginess (or extravagance), rowdiness, bad language, sexual promiscuity, arrogance and just plain rudeness, and impatience with local customs and sense of time are just a few of the characteristics some Western travellers have become known for.

Some of that behaviour occurs because travellers are not prepared to encounter a totally different culture. They may behave badly out of ignorance or because they're overwhelmed and confused – suffering from "culture shock." Preparing yourself by researching the country, its history, culture, and religious beliefs will make the travel experience easier for you and your hosts.

✳✳✳✳✳✳✳✳✳✳✳✳✳✳✳✳✳✳✳✳✳✳✳✳✳✳✳✳✳✳✳✳✳✳✳✳✳✳✳

Travelling closer to the ground is not only cheaper, it's more stimulating.

The Ecumenical Coalition on Third World Tourism recommends the following code of ethics for travellers:

▬ Travel in a spirit of humility and with a genuine desire to meet and talk with local people.

▬ Be aware of the feelings of the local people; prevent what might be offensive behaviour.

▬ Cultivate the gift of listening and observing, rather than merely hearing or seeing, or thinking you know all the answers.

▬ Realize that other people may have concepts of time and thought patterns that are very different – not inferior, only different.

▬ Instead of seeing the beach paradise only, discover the richness of another culture and way of life.

▬ Get acquainted with local customs; respect them.

▬ Remember that you are only one among many visitors; do not expect special privileges.

▬ When bargaining in the shops, remember that the poorest merchant will give up a profit rather than give up his or her personal dignity.

▬ Make no promises to local people or to new friends that you cannot deliver on.

▬ Spend time each day reflecting on your experiences in order to deepen your understanding. What enriches you may be robbing others.

▬ If you want a home away from home, why travel?

Staying in local inns is not always an option for everyone – using a squat toilet is no fun if you have arthritis. Here are some suggestions from the Center for Responsible Tourism that you can still follow:

▬ If you travel in a group, keep it small.

▬ Allow plenty of time in each place. It's better to visit a few places leisurely than several places in a rush.

▬ Plan visits to local markets rather than the tourist shops.

▬ Try to eat local foods rather than retreating to international cuisine in hotels and restaurants.

▬ Wherever and whenever possible, use local transport.

▬ Try to spend as much time in rural areas as in cities.

* *

Don't assume, by the way, that eating in local restaurants puts you in danger of food poisoning. Some experienced Third World travellers insist that it's actually safer to eat in moderately priced local restaurants. However, depending on the area, there are particular types of food you should avoid. For more information you can consult one of the many guides to healthy travelling, such as *Travelling in Tropical Countries* by Jacques Hébert (Hurtig, 1986) or *How to Stay Healthy Abroad* by Richard Dawood (Oxford, 1989).

Other ways to get to know a country without exploiting it include studying a language or some other subject or helping in a development project. *The Directory of Alternative Travel Resources* lists hundreds of such options around the world (**One World Family Travel Network**, P.O. Box 3417, Berkeley, CA, 94703, USA).

The burgeoning business of travel has become a threat not only to local cultures and economies but also to the environment. Thanks to tourism, parks are created and cultural monuments preserved that might otherwise be destroyed. But the trend of easy access to world-famous but environmentally sensitive sites recently caused Sir Edmund Hillary to comment that Mount Everest is "in danger of becoming a rubbish dump." A stampede of camera safaris to "capture" wildlife in its natural habitat has put increasing pressure on many of the world's most singular and treasured ecosystems.

The ecologically fragile Galapagos Islands, where Darwin uncovered many of the basic rules of evolution, is evolving into a "must see" world tourist attraction. Although a yearly limit of 12,000 visitors had been set, this was quickly upped to 25,000 to take advantage of tourist dollars. By 1986, visitors numbered in excess of 30,000.

In Nepal, where locals rely on wood for fuel, the influx of trekkers has resulted in serious deforestation, as trees are burned to provide visitors with heat, baths, boiled water for drinking, and Western-style meals that take longer to cook. Deforestation has resulted in loss of topsoil, dangerous landslides, and the disappearance of local wildlife.

On tropical islands, the clearing of swamps to build tourist facilities increases soil erosion. The soil washes onto the coral reefs, killing the coral; without the coral reefs to act as breakwaters, the local shorelines are eroded, and the beaches that attracted the tourists in the first place are washed away. The tourists move on, but the local ecosystem is destroyed.

Mount Everest is in danger of becoming a rubbish dump.

However, there are tour companies that have made an effort to limit and even counteract the environmental damage. For example, **Worldwide Adventures** was the first Western tour company to carry kerosene on its Nepalese treks as an alternative to burning wood. The company, which offers everything from a trip up the Amazon to a sail around grizzly-bear birthing grounds, also invests in conservation projects in the areas it visits and gives a thorough briefing on environmental and cultural do's and don'ts to its clients. If you're considering a travel package, check to see that the tour company has a sound environmental policy.

Tourists also cause environmental problems in their shopping. Although the United Nations Convention on International Trade in Endangered Species (CITES) prohibits the export of endangered animal and plant life, that doesn't mean you won't find those items for sale in tourist markets and shops. Inform yourself about species that are on the endangered list in your destination. Many airports have displays about CITES, and you can also get information from the **Canadian Wildlife Service**, Environment Canada, Ottawa, ON, K1A 0H3.

****

The Green Tourist does not buy:

- **Jewellery made from coral**
- **Combs and ornaments made from the shells of endangered turtles**
- **Live plants and animals**
- **Ivory products**
- **Pictures made from bird feathers**
- **Rhino horn, leopard skins, or other animal souvenirs**

The United Nations Convention on International Trade in Endangered Species (CITES) prohibits the export of endangered animal and plant life. But you'll still find many of these items for sale in tourist markets and shops. Inform yourself about species that are on the endangered list; many airports have displays about CITES. You can also get information from the Canadian Wildlife Service, Environment Canada, Ottawa, ON K1A 0H3.

Travelling AT HOME

Prime destinations for both Canadian travellers and overseas visitors are Canada's 33 national parks and other wilderness areas. More than 21 million people visited them in 1987.

Unfortunately, an increase in visitors is happening just as governments are cutting park funding. That puts more responsibility on all of us as individual visitors to make sure our parks are maintained as natural paradises where humans can continue to seek refuge in harmonious co-existence with plants and animals. The important word to remember is "low-impact"; when you visit a park, whether you're just stopping for a few hours or staying for several days, no one should see any evidence that you've been there.

The Woodsman's Code, developed by the Conservation Council of Ontario and the Canadian Camping Association, tells you how to be a "low-impact" user of our parks and wilderness areas. It was written for campers, but many of the rules apply equally to hikers and other daytime visitors.

Planning

1 Keep the group size small to reduce the impact on campsites.

2 Prepare carefully:

▬ Research the trip: become familiar with the geography of the area and learn the local fishing, forestry, and wildlife laws and regulations.

▬ Discuss the trip thoroughly with all group members, including emergency plans. Become familiar with the Woodsman's Code.

▬ Make sure to carry enough food for the whole trip. Don't rely on the environment for emergency food.

▬ Carry proper equipment, including good shelter, clothing, rain gear, etc.

Travelling

3 Stay on existing trails and portages. Following wildlife trails is better than cutting new trails.

4 Use switchbacks on trails. Don't cut a new trail just to save a few metres of walking; you might be creating a severe erosion problem. Walk *through* puddles to avoid widening trails.

5 Don't wear ridge-soled boots unless it's absolutely necessary. These boots destroy moss cover and other vegetation, which leads to erosion. Running shoes are much better.

Campsites and Shelters

6 Use existing campsites. Keep your heavy-use areas small, to minimize soil compaction. Very few plants can grow back where soil has been compacted. Don't expand the campsite.

7 If you have to choose a new campsite, choose one that will disturb trees or shrubs as little as possible. If your tent is pitched on top of live vegetation, move it after one or two days.

8 Don't "improve" the site by pulling vegetation, building walls, or digging trenches around your tent.

9 Use a floored tent. Don't make shelters or bedding out of branches or other natural material. A lightweight plastic tarp with ropes can provide extra protection from wind and rain.

Fire

10 Use camp stoves whenever possible. The new ones are convenient, reliable, and easy to carry. This will reduce fire hazards, save wood, and give you a chance to do some stargazing!

11 If a fire is necessary, keep it small and use existing fire pits.

12 Where there is no fire pit, build your fire on bedrock or pure sand, preferably close to water. Where this is not possible, dig a pit down to the mineral (clay) layer of the soil. Pick a spot that avoids roots, overhanging trees, needles, leaves, and other forest litter.

13 Use only dead wood for the fire.

14 Drown your fire thoroughly with water. Stir the ashes. Add more water. Leave unused firewood for the next campers. Pack out all the unburned bits of garbage, such as aluminum foil, cans, etc. In wilderness sites, eliminate all traces of the campfire.

Smoking

15 Sit down to have a smoke, so you can be careful with your ashes.

16 Remember that cigarette butts are garbage and should be packed out. Filter tips won't degrade naturally. A small metal cough-drops box is a good way of storing them. You can count them at the end of the trip!

Human Waste

17 Use existing outhouses or latrines wherever possible.

18 Where necessary, dig a small hole 15 to 20 cm deep in soil, at least 35 m from any open water. Soil can act as a filter for nutrients and bacteria, and can protect lakes and streams from pollution by human waste. Use single-ply undyed toilet paper, and bury everything completely.

Other Waste

19 Carry out everything you carried in! The rule is: BURN IT OR BASH IT, THEN BAG IT, *AND BRING IT BACK*.

20 Wash dishes, clothes, and yourself in a dishpan, *not* in the lake or stream. Rinse away from open water. Dump waste water in a hole located at least 35 m from open water. Use liquid or bar soaps instead of detergents.

21 Fish guts attract swarms of flies, which can ruin a campsite for following users. After cleaning and gutting fish, collect the waste and bury it in soil far from the campsite. You can also paddle it out to a rock island, where birds will eat it.

Wildlife and Natural Food

22 Remember, you are a guest in someone else's home. Avoid disturbing wildlife, especially young animals or nesting birds.

23 Avoid overfishing and overhunting. Remember that all plants and animals are protected in provincial parks.

24 Obey all fish, game, and forestry regulations. Talk to a conservation officer about them if you have questions.

25 Avoid picking edible wild foods except where they are abundant. Wildlife depends on this food. Collect wild mushrooms only if you are an expert: many common mushrooms are deadly poisonous.

26 Never feed wildlife: this interferes with their natural habits, and you may be bitten by a nervous or rabid animal. Guard your food and garbage carefully from wildlife, and keep all food out of your tent. In bear country, seal all food and garbage in your pack overnight, and hang your pack at least 6 m high on a rope between two trees. Bears that develop a taste for handouts eventually have to be shot.

In bear country, seal all food and garbage in your pack overnight, and hang your pack at least 6 m high on a rope between two trees.

Clean Up Others' Mistakes

27 Pack out any garbage that you find.

28 Cover up fire pits and latrines that are in poor locations.

29 Tell the local parks or natural resources officials about any major problems that you find.

Certain parks provide organized sports, such as horseback riding, guided nature tours, rafting, and scuba diving. For further information about facilities, activities, and rules of the various national parks, contact the Inquiry Centre of **Environment Canada** (Hull, PQ, K1A 0H3). For information on provincial parks, get in touch with the appropriate ministry in your province.

Not all "parks" are wilderness areas. Several of Canada's national and provincial parks, such as L'Anse-aux-Meadows, location of a Viking settlement, and Fort Louisbourg, are important historic sites, and certain river systems have been designated Canadian Heritage Rivers. These should be treated with the same care and respect as wilderness parks. Information on historic parks is also available from Environment Canada. To find out more about the Heritage River program, contact **Canadian Heritage Rivers**, Parks Canada, Ottawa, ON, K1A 1G2.

cottaging

There are roughly 568,000 cottages in Canada. No one has a better chance to observe the effects of pollution than cottagers, who return to the same bit of nature every year. But the causes of pollution are varied and sometimes complicated, and many cottagers are unsure how they can avoid contributing to the problem.

In cottage areas, pollutants and raw waste are reaching the surface water and groundwater. The pollutants interfere with the ecosystems of lakes and rivers. The raw and untreated waste introduces unwanted nutrients into the water, which causes excess growth of algae and imbalance in the ecosystem and the food chain. Beer cans tossed overboard, motor oil and antifreeze spills from powerboats, shampoo and soaps, dishwasher detergent – all contribute to the growing mess. In addition, human wastes can transmit viruses such as hepatitis, polio, and intestinal flu.

In order to avoid contributing to the problem, consult the information on environment-friendly household products in Chapter 3 and on gardening and pesticides in Chapter 6. Since cottages are almost invariably close to water, it's even more important to avoid using polluting chemicals there. If you have a recycling program at home, bring back the recyclables from your cottage garbage.

Since your cottage is probably not attached to a municipal sewage system, there's also the question of how you're going to dispose of wastes. One common solution is the septic tank.

The septic tank holds about 2,700 to 4,500 L of human waste underground. Water and lime stimulate the production of bacteria within the waste solution, starting the process of decomposition. The result, called "septage," is filtered and leached through a tile or gravel bed and absorbed into the ground.

Some septic tanks do not use water. Because much of the waste remains as solids in the tank, these must be inspected for clogging and pumped out regularly, and the waste taken to municipal treatment sites. The sludge can also be composted to produce rich fertilizer.

However, septic tanks are not the perfect solution. They should not be used in areas that experience high fluctuations in the water table. This increases the chance of untreated waste entering the surface water or groundwater.

An alternative to septic tanks for human waste is the biological toilet. Body wastes and toilet tissue are turned into water by the continuous action of enzymes and bacteria. The recycled waste becomes a clear and odourless liquid containing no pathogenic organisms. There is no residue, no sludge to be removed. Once a week a package of freeze-dried bacteria and enzymes is added to the toilet, and every two years the charcoal filters must be replaced.

BUG REPELLENT

At best, black flies, mosquitoes, and other biting insects are an irritation. At worst, they can cause serious illnesses, such as meningitis, Rocky Mountain fever, and Lyme Disease. However, some bug repellents contain dangerous chemicals. Fortunately, you can protect yourself to a certain extent by wearing proper clothing and using natural repellents.

Clothing should be loose and light-coloured, covering as much exposed skin as possible. Watch out for small openings, and keep collars and shirt and pant cuffs tightly closed. If you must use repellents, at least you can restrict them to the hands and face by keeping yourself well-covered.

A natural bug repellent is citronella, the oil extracted from citrus fruits, which is available at drugstores.

Some people believe that you can also eat your way to protection by consuming lots of garlic (mosquitoes supposedly don't like the smell any better than we do), avoiding refined sugar and flour, and taking extra B vitamins or brewer's yeast.

Once bitten, you can relieve the itching by rubbing the bite with raw garlic, lime or lemon juice, damp salt, or Vitamin C powder.

Beer cans tossed overboard, motor oil and antifreeze spills from powerboats, shampoo and soaps, detergents – all contribute to the mess.

PROVINCIAL MINISTRIES of TOURISM and RECREATION

are the most comprehensive sources of information on leisure activities across Canada. You can also check with conservation groups, many of which publish their own newsletters and magazines. See page 156 for some addresses.

All-Terrain Vehicles

This sport can cause serious damage to the ecosystems that are favourite spots for ATV drivers – shallow rivers, gravel bars, stream beds, sand dunes, and deserts. ATVs can scar the land, kill wildlife, and destroy vegetation, creating noise pollution as they go.

* * * * * * * * * * * * * * * * *

Aquariums

Aquariums can be fun and educational, and many promote the conservation message. However, animal rights activists have some serious concerns about their treatment of marine animals: whales, porpoises, dolphins, and others suffer undue stress during their capture; they may be kept in holding tanks that are far too small; many are seasonally transported from one aquarium to another.

Studies indicate that many marine mammals held in captivity have a reduced lifespan and live a poorer-quality life. Lax Canadian animal protection regulations have also encouraged a number of foreign aquariums to come into Canada's territorial waters to acquire marine mammals for their collections.

* * * * * * * * * * * * * * * * *

Beaches

There's already enough garbage in our water and washing up on our shores without tourists adding to it, so don't litter when you go to the beach. Garbage on the beach has always been considered an eyesore, but recently environmental workers have recognized that it's a serious threat to marine life. Plastic is a particular problem; seals, seabirds, marine turtles, and other wildlife, some of them endangered species, have died from eating plastic or becoming entangled in it. Sea turtles, for instance, eat plastic bags because they mistake them for jellyfish, which are among their natural prey.

* * * * * * * * * * * * * * * * *

Birdwatching

Check with your provincial tourism ministry or naturalists' society for bird-watching opportunities. While this is an easy and enjoyable solitary activity, there are also birdwatching clubs, competitions, and tours.

* * * * * * * * * * * * * * * * *

Canoeing, Kayaking, Sailing, Windsurfing

All these boating activities have the advantage of using little or no fuel. When using large sailboats with motors and toilets, be careful not to spill any fuel, oil, or sewage while on the water.

* * * * * * * * * * * * * * * * *

City Farms

Many urban children grow up with no experience of animals other than household pets, and little idea of where eggs, milk, and other animal foods come from. Fortunately, some cities have petting zoos and farms where urban dwellers can get to know domestic animals.

Be aware of toxicity levels in the area where you are fishing.

Cycling

Environmentally speaking, this is one of the most respectable leisure activities around. As well as cycling at home, you can take cycle vacations abroad, and outfitters in vacation spots across Canada will rent cycles by the day or hour. Follow the road rules, however, and don't ride in heavy traffic unless you're experienced. Careless cyclists in some major cities have caused accidents and are beginning to give the sport a bad name.

****

Farm Holidays

Staying on a farm is another way of enjoying nature and educating the family. While participation is not mandatory, many farms encourage guests to get involved in farming activities. You can get a list of farm accommodations from your provincial ministry of tourism.

****

Field Studies

As well as the outdoor activities that come easily to mind, there are a host of more unusual possibilities that you can learn about from your provincial ministry, such as maple sugaring, seal-watching, and guided tours of woodlots. As an alternative, the Foundation for Field Research links volunteers from the public with scientists whose projects are funded by the foundation and who need research assistance. Expeditions can last anywhere from a weekend to a month. For more information, write to the **Foundation for Field Research**, P.O. Box 2010, Alpine, CA 92001-0020, USA.

Fishing

According to the Angler's Code, created by the Ontario Ministry of Natural Resources, a good angler:

■ respects private property and the rights of others

■ knows and obeys the fishing regulations

■ does not damage fish habitat

■ puts safety first in the use of his equipment and the enjoyment of the sport

■ takes pride in his skill

■ helps others to understand the recreation of fishing

■ leaves the environment as clean as he found it; he does not litter.

A good angler has respect for his quarry, before and after catching it, and knows there is much more to fishing than taking his limit. Be sure to follow fishing regulations in your province and be aware of toxicity levels in the area where you're fishing.

****

Giardiasis

Contrary to popular belief, Montezuma's revenge is not restricted to foreign climes. Canadians can contract giardiasis – also known as "beaver fever" – from contact with human or animal feces or drinking water contaminated by feces. The giardia parasite lodges in the digestive tract, causing diarrhea, vomiting, and other flu-like symptoms. Giardiasis can be treated, but prevention is easier. Since the parasite tends to be found in lakes, ponds, and inadequately treated water, campers and cottagers should be particularly careful. If you're using untreated water, boil it for at least five minutes. Don't swim in beaver ponds, and keep your own waste well clear of water.

Hostelling

There are more than 5,000 hostels worldwide offering cheap accommodation in everything from an ancient jail to a sailing ship. The Canadian Hostelling Association has 60 hostels that offer year-round accommodation, outdoor recreational programs led by environment-conscious volunteers, and travel talks and films. To stay in a hostel, you'll need a membership, which also entitles you to budget travel information and a bi-monthly newsletter. Students get a Student ID card entitling them to discounts on admission to museums, galleries, and events around the world. For information on Canadian and international hostelling, write to the **Canadian Hostelling Association**, 1600 James Naismith Drive, Suite 608, Gloucester, ON, K1B 5N4; (613) 748-5638.

✳ ✳ ✳ ✳ ✳ ✳ ✳ ✳ ✳ ✳ ✳ ✳ ✳ ✳ ✳

Houseboats

Although floating down lakes and rivers on a houseboat sounds like the perfect natural vacation, critics are concerned about "grey water," the discharge from sinks and showers that all too often ends up in said lakes and rivers. Houseboat discharges are not currently regulated and may contain concentrations of fecal and food bacteria as well as phosphates from detergents.

✳ ✳ ✳ ✳ ✳ ✳ ✳ ✳ ✳ ✳ ✳ ✳ ✳ ✳ ✳

Lead Weights and Shot

Anglers should never use lead lures and weights, and hunters should use steel rather than lead shot, to avoid poisoning birds and fish. A single pellet can kill a bird, and lead poisoning is not a pleasant way to die. The problem is compounded when poisoned animals are eaten by predators.

Currently about 3,600 tonnes of lead ammunition falls on North America's fields and wetlands every year. It's difficult to estimate just how much wildlife is affected by this lead, but the U.S. Fish and Wildlife Service has put the number of birds alone at 1.5 to 3 million in the U.S.

Lead shot has been banned in certain heavily hunted areas of Canada. By the fall of 1990, Ottawa had designated part of Ontario's Lake St. Clair and two regions in British Columbia as "non-toxic shoot hunting zones," under the migratory bird hunting regulations. Environment Canada is currently considering placing further controls on the use of lead shot in other problem areas by 1992.

✳ ✳ ✳ ✳ ✳ ✳ ✳ ✳ ✳ ✳ ✳ ✳ ✳ ✳ ✳ ✳ ✳

Four-legged family members may enjoy a vacation as much as the rest of us, but don't let them run wild in natural areas, where they may injure or be injured by the local animals. Make sure your pet has all its shots, and if you're planning to travel outside the province or country, have the documentation to prove it. Find out the location of the veterinary clinic closest to your destination, in case of emergencies.

The Green Consumer is, of course, a responsible pet owner. Dogs and cats should be spayed or neutered. "Poop" should always be "scooped." Pets should never be left out in the sun in a locked car with the windows rolled up.

The Canadian Hostelling Association has 60 hostels that offer year-round accommodation.

"GREEN ROVER"

A GREEN DEVELOPMENT

Most photofinishing laboratories recover silver from their wastes before they are discharged to the sewers. The labs can easily install special equipment in their plants to recover the silver on-site, or save their waste solutions for recycling companies who will recover the valuable metal for them.

In addition, most large photofinishing laboratories have installed good pollution control and resource recovery equipment — many regenerate their used bleach and fixer, a few even recycle their developing fluid, in addition to the silver in the waste water. Along the way, they've also found that recycling contributes to company profits. If you're willing to wait a few days to get your pictures back, it's likely they'll be sent to a large centralized lab that practises sound waste management.

Many photofinishers also recover plastic materials, such as black or clear plastic film canisters. The collected plastics can be recycled into other products, such as building board. Some municipalities will also collect film containers in their curbside or Blue Box recycling programs.

However, a number of small photo labs have not adopted all of these recycling programs — almost one-third of these labs do not even recover their silver. The waste water from many so-called mini-labs goes directly into the municipal sewer system. So ask your photoprocessing shop whether they recycle their chemicals and reclaim their silver. If they don't, take your film elsewhere.

Photography

For the most part, photography is a wonderful way to enjoy nature without doing any damage. But have regard for the feelings of your subject: both people and animals can be annoyed by the over-eager photographer. Owls, for example, are confused and frightened by the flashes used by nighttime photographers; bright flash bulbs can even harm their sensitive eye tissue. Paying too much attention to any bird of prey can disrupt essential hunting and feeding patterns. And paths photographers make to nests or dens can also lead predators to those same sites.

* * * * * * * * * * * * * * * * * *

Powerboats

Though powerboats are an important method of transportation for many Canadians, they do contribute to the pollution of our waters. The Allied Boating Association cites three major problem areas in powerboat use:

▬ Leaded gasoline: the association has been pushing for its abolition.

▬ Anti-fouling paint, used on the bottoms and sides of boats to protect them from marine life such as algae and barnacles. This paint usually contains tributyl tin (TBT), which has been called the most toxic substance ever introduced into the aquatic environment. It may also be used in fungicides used on air conditioners, pipes, and fishnets, and has been banned in many jurisdictions, including France and the U.K. As an alternative, you may be able to find a non-toxic wax, which does not kill the algae but instead inhibits their ability to stick to boats.

▬ Maintenance: Some boat owners start their motors in the lake for the first time each summer to clean them out and get them prepped for summer use. This often results in blowing antifreeze and engine residue into the water.

The association recommends that all maintenance work be done on shore and that you use "plumber's antifreeze," which is available at hardware stores and is less toxic.

* * * * * * * * * * * * * * * * * *

Rail Travel

Rail passes for Canada and Europe can provide you with an ecologically sound and relatively cheap way to travel. Eurail passes must be purchased before you leave Canada, through the Canadian Hostelling Association or by writing to **Eurail Pass Distribution Centre**, Box 300, Succursale R, Montreal, PQ, H2S 3K9. There are also passes available for some individual countries. Canadian passes can be purchased from the Canadian Hostelling Association or from any VIA agent.

* * * * * * * * * * * * * * * * * *

Safaris

Although poachers continue to threaten the survival of many species, cameras have taken the place of guns for many tourists. But even they can sometimes be a problem when placed in the hands of over-eager safari-goers. After rescuing a pride of lions that had been surrounded by tourists in jeeps, one exasperated park warden in Kenya suggested that if tourists were going to take liberties with the lions, maybe the lions should be allowed to take liberties with the tourists. If you plan to take a safari, make sure your tour company has a respectful attitude towards wildlife.

* * * * * * * * * * * * * * * * * *

Skiing

Downhill skiing may seem to be passive recreation, but the creation and clearing of slopes can be damaging to mountain ecosystems. Cross-country skiing or snowshoeing is a good alternative; it's best to stay on marked trails, for your own safety and that of the environment.

Snowmobiling

A brilliant Canadian invention, the snowmobile has made life and work easier and safer for people living in areas with heavy snowfalls. However, snowmobilers should stick to existing trails whenever possible, to avoid damaging the environment and frightening wildlife. For more information on using your snowmobile safely and wisely, write to the **Canadian Council of Snowmobile Organizations**, 98 Marshall Street, Barrie, ON, L4N 4L5.

* * * * * * * * * * * * * * * * * *

Whalewatching

Whalewatching trips that range from one-day cruises to week-long study trips have become an effective way of raising concern about preserving the whales. They have also become a $12-million industry, and observers are now concerned that the number of whalewatchers may be having a harmful effect on the whales, driving them away from their natural feeding grounds. Guidelines for boat observation are needed to protect both the whales and this fledgling industry. In the meantime, some whalewatching in the Gulf of St. Lawrence and on the West Coast can be done from shoreline observation posts.

* * * * * * * * * * * * * * * * * *

Wildlife

Nine out of ten Canadians take part in some form of wildlife recreation, such as birdwatching or photography; about 3.6 million every year take trips specifically for that purpose. Despite our collective concern, however, there are currently 147 species of wild plants and animals on the endangered list, and 16 species that no longer exist in Canada.

Young animals may sometimes appear to have been abandoned by their mothers, but in most cases they have not, so it's best to leave them alone. In most cases if a baby bird has fallen from its nest, it is best to place it back under cover, let the parent take care of it — and keep the cat indoors. It is a myth that an animal will abandon its young if it has been handled by humans. Your library or local humane society will have more information on caring for orphaned or injured wildlife.

* * * * * * * * * * * * * * * * * *

Zoos

Canada's zoos, wildlife displays, circuses, and animal collections are largely unmonitored, unregulated, and uncontrolled. Animal protectionists have raised serious questions about zoos. Capturing the specimens causes considerable stress, and the animals may be killed or injured in the process. Once an animal is removed from its natural habitat, especially if it is caged, its behaviour changes. It may show signs of psychological damage, such as pacing, attacking other animals, and self-mutilation. It may suffer physical problems and be exposed to disease. Its life expectancy may decrease.

Some zoos are definitely better than others. Obviously animals should be well fed, and should be given good veterinary care, "toys" and other diversions, socialization opportunities, protection from the elements, and as much space as is appropriate (including some "private space" where they can escape from public view). But good zoos also spend sizeable proportions of their budgets on public education, research, preservation of animals' natural habitats, and re-releasing programs. Only about 20 zoos (out of an estimated 500) have registered as accredited members of the Canadian Association of Zoological Parks and Aquariums (CAZPA) and met their minimum standards.

Zoocheck Canada is a charitable organization established to protect captive animals in zoos and wildlife displays throughout Canada, and to participate in conservation projects of direct benefit to animals in the wild. It documents zoo conditions and is pushing for implementation of a regulated licensing system. For more information, contact Zoocheck Canada, 5334 Yonge St., Suite 1830, Toronto, ON M2N 6M2; (416) 696-0241.

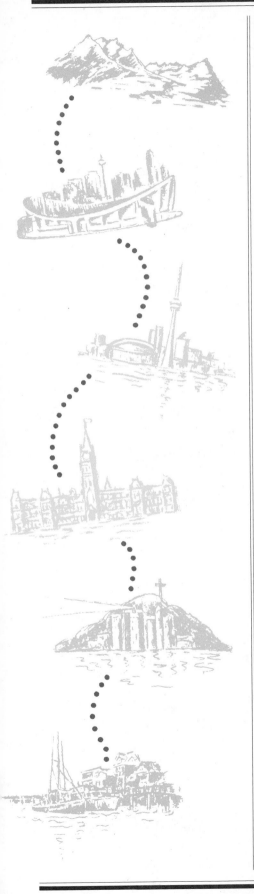

FOR MORE INFORMATION

Many of these groups have provincial affiliates.

■ **Canadian Association of Nature and Outdoor Photographers**
19 Mercer Street
Suite 301
Toronto, ON
M5V 1H2

■ **Canadian Camping Association**
1806 Avenue Road
Suite 2
Toronto, ON
M5M 3Z1

■ **Canadian Nature Federation**
(publisher of *Nature Canada* magazine)
453 Sussex Drive
Ottawa, ON
K1N 6Z4

■ **Canadian Parks and Recreation Association**
333 River Road
Vanier, ON
K1L 8B9

■ **Canadian Parks and Wilderness Society**
(publisher of *Borealis* magazine)
160 Bloor Street East
Suite 1150
Toronto, ON
M4W 1B9

■ **Canadian Wildlife Federation**
1673 Carling Avenue
Ottawa, ON
K2A 1C4

■ **Wilderness Canoe Association**
Box 496
Station K
Toronto, ON
M4P 2G9

■ **World Wildlife Fund**
60 St. Clair Avenue East
Toronto, ON
M4T 1N5

****

PROVINCIAL TOURISM DEPARTMENTS

Travel Alberta Box 2500 Edmonton, AB T5J 2Z4	**TravelArctic** Yellowknife, NT X1A 2L9	**Tourism British Columbia** Parliament Buildings Victoria, BC V8V 1X4	**Travel Manitoba** 7th floor 155 Carlton Street Winnipeg, MB R3C 3H8
Tourism New Brunswick P.O. Box 12345 Fredericton, NB E3B 5C3	**Tourism Branch Newfoundland Department of Development and Tourism** P.O. Box 2016 St. John's, NF A1C 5R8	**Nova Scotia Department of Tourism and Culture** P.O. Box 456 Halifax, NS B3J 2R5	**Ontario Ministry of Tourism and Recreation** Queen's Park Toronto, ON M7A 2E5
Prince Edward Island Visitor Services P.O. Box 940 Charlottetown, PE C1A 7M5	**Tourisme Québec** P.O. Box 20,000 Quebec City, PQ G1K 7X2	**Tourism Saskatchewan** 1919 Saskatchewan Drive Regina, SK S4P 3V7	**Tourism Yukon** P.O. Box 2703 Whitehorse, YT Y1A 2C6

THE CANADIAN GREEN CONSUMER GUIDE DIRECTORY

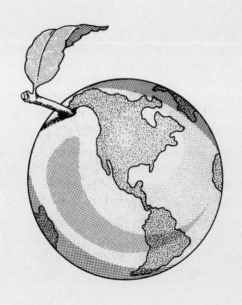

11. THE GREEN DIRECTORY

A listing in The Green Directory does not imply a recommendation or endorsement by the Pollution Probe Foundation. Until governments pass more stringent regulations on "green" products, the Green Consumer must be prepared to ask questions about the contents of products, the type of packaging used, and how environment-friendly and recyclable the products and packaging are. The listings here are meant only to give you a place to start looking.

Does Buying Green Mean Paying More?

Not necessarily. Some environment-friendly products cost more than their conventional counterparts; that's partly because retailers charge what the market will bear, and consumers have generally been willing to pay more for green items. Another factor is that manufacturers often need to make large capital investments to get a new product line going. But one of the key goals of being a Green Consumer is to reduce overall consumption, so Green Consumers could end up paying less simply because they buy less.

THE GREEN DIRECTORY

Introduction

Buying power is the driving force behind Green Consumerism. It is the consumer's ability to change from Brand X to Brand Y — or even to stop buying a particular product altogether — that makes producers sit up and take notice. This book provides the information that will allow Green Consumers to use this power most effectively for the benefit of the environment.

It isn't always easy to point to the "best" choices. Environmental science is a new area, and we just don't know yet how to tackle some of the problems we've created. For instance, the lack of contents labelling makes it impossible for consumers to tell which products are truly environment-friendly. Because some companies consider this information "proprietary" — a trade secret — most of our non-food products have no ingredient labelling.

The Green Consumer also faces a number of frustrating choices where all the options are imperfect. Is cotton clothing better than polyester, because artificial fibres use up non-renewable resources and go through a lot of processing, which creates pollution? Or is polyester better after all, because cotton growers use massive amounts of chemical fertilizers and pesticides, which run off into local water supplies? There is no definitive answer here. For the time being, the rule of thumb must be: When in doubt, choose the natural product over the synthetic.

Green Consumers need to make their voices heard directly, too. Write to politicians to demand better labelling, stricter regulation and enforcement, and stronger commitment to cutting pollution. Write to manufacturers and retailers to ask for more green products. When you decide to stop buying one brand and switch to another, tell both makers why. Businesses *are* interested in your behaviour. That's why they spend thousands trying to decipher it through surveys and studies.

You may wonder whether you can really make a difference, given the overwhelming scope and complexity of the dangers our planet faces. The answer is yes, because the interconnectedness of our ecosystem means that every action you take starts a chain reaction — for good or for ill.

So don't underestimate your power and influence. With this Green Directory in your hands, you can make a difference at the supermarket, at the hardware store, at your workplace, when you pay your bills and your taxes — in short, whenever and wherever you spend your money. Responsible shopping need not cost the earth.

How We Put
The Green Directory Together

There were a lot of "green products" described in the first edition of *The Canadian Green Consumer Guide*. These were used as examples, to show what alternatives were available or to illustrate that some progressive companies were beginning to respond to the growing consumer demand for environmentally friendlier products.

Although Pollution Probe is constantly collecting information on green product developments, it had not attempted to systematically investigate the wide array of green products available. This meant that in certain areas the first edition of the book overlooked some very good alternative products. And in the 18 months since the guide was published, a great many more environment-friendly goods and services have come on the market.

For the second edition, Pollution Probe decided to canvass the business community, looking for products and services "that do not harm the environment as a result of their manufacture, use or disposal." These firms could then be incorporated in a new chapter — The Green Directory — as an aid to consumers looking for green alternatives. Several retail chains, Loblaw Companies Ltd. being one high profile example, are making space on their shelves for a number of new, green products. This is a trend many environmentalists applaud and support. For the sake of space, Probe has not listed these stores in The Green Directory. Nor have we included the hundreds of health food stores across Canada which carry environment-friendly products. However, green consumers are heartily encouraged to investigate both these sources. Probe identified a number of product and service categories and drafted a list of criteria that companies had to comply with to be included in The Green Directory. Companies were asked to fill in the application form only if they could answer "Yes" to all of the criteria questions. To be accepted the form had to be signed by a senior company official (preferably the president), witnessed, and dated.

Pollution Probe searched for prospective firms in its own files, association and trade directories, Yellow Page listings, magazine advertisements, brochures picked up at environmental conferences and trade shows, and leads passed on by the book's review committee. It also sent out notices to trade journals, environmental magazines and the business press in Canada's ten or twelve largest cities.

In all Pollution Probe compiled well over 1400 prospective companies that offered environmental goods or services in Canada. Each of these firms received an application form. Several hundred received follow-up fax transmissions or telephone calls to clarify or complete the information in their applications. *No company paid to have its products or services listed in The Green Directory*.

And so The Green Directory was born. It is anticipated that as future editions of *The Canadian Green Consumer Guide* are published, The Green Directory will grow and grow. If your firm wants to be considered for future listings, please write to us and you'll receive an application form when the next edition is being prepared.

The Green Directory, Pollution Probe Foundation, 12 Madison Avenue, Toronto, ON M5R 2S1

Note to Consumers

We have listed the criteria to which a company's product and/or services had to comply in order to be listed here in The Green Directory. If, in your own experience, you find that company does not live up to its claims, or if you have problems in dealing with the company, let us know. Write to: The Green Directory, Pollution Probe, 12 Madison Avenue, Toronto, ON M5R 2S1.

Criteria For Green Product Companies

Yes *No*

☐ ☐ Have you incorporated (waste) reduction, reuse and recycling principles into your manufacturing processes?

☐ ☐ Can you assure that the health of humans, other animals and plantlife is not endangered in any way due to the manufacture, use and disposal of your product?

☐ ☐ Will you supply consumers with information on how to safely handle, use and dispose of your product?

☐ ☐ Will you supply consumers with a complete list of your product's ingredients upon request?

☐ ☐ Is your product cruelty-free (neither the product nor its ingredients have been tested on animals)?

☐ ☐ Is your product packaged in a refillable container when possible? If not appropriate, is the container recyclable?

☐ ☐ Is your product available in bulk form (that is, without packaging) when appropriate?

☐ ☐ Will your company take back shipping materials for recycling or reuse?

☐ ☐ Will you accept the return of unused portions of your product for reuse, recycling or disposal?

☐ ☐ Is the product available in Canada?

****

Criteria For Green Service Companies

Yes *No*

☐ ☐ Have you incorporated waste reduction, reuse and recycling principles into your business operation?

☐ ☐ Can you assure that the health of humans, other animals and plantlife is not endangered in any way as a result of the services you offer?

☐ ☐ Will you supply consumers upon request with a complete list of the chemical products used in your service?

☐ ☐ Will you supply consumers and staff with information on how to safely handle, use and dispose of the products used in your business?

☐ ☐ Is your service cruelty-free (neither the products used in your service nor their ingredients have been tested on animals)?

☐ ☐ Is your business in compliance with all applicable environmental regulations and other requirements?

☐ ☐ Is the service available in Canada?

THE GREEN DIRECTORY

* * * * * * * * * * * * * * * * * *

PRODUCTS

If the product brand name is the same as the company name, it is not repeated. If the brand name differs from the company name, it follows the product description.

Cleaning and Homecare

There are instances when a home remedy won't do the job and you need to turn to a commercial cleaning product. When you do, be sure to choose products that are readily biodegradable and that contain the fewest synthetic chemical additives. Look for products made of natural ingredients, that are not tested on animals, and that come in refillable or recyclable packaging.

Aerokure
P.O. Box 22
Sherbrooke, PQ J1H 5H5
(819) 821-2238
odour eliminator

Amway
399 Exeter Rd.
London, ON N6A 4S5
(519) 685-7700
cleaners; laundry products

Caeran
25 Penny Lane
Brantford, ON N3R 5Y5
(519) 751-0513
cleaners; laundry products

Dynamic Essentials
3 Waterloo St.
New Hamburg, ON N0B 2G0
(519) 662-2520
cleaners; furniture polish; laundry products

Earth-Wise
R.R. 2
Roslin, ON K0K 2Y0
(613) 477-2693
air fresheners

EMJ Environmental Products
R.R. 6, Highway 24 South
Guelph, ON N1H 6J3
(519) 837-2444
cleaners (Brown Bag, Simply Clean); laundry products (Simply Clean)

Gerry Brushett Enterprises
Comp. #27, Mayflower Ave.
RR #1
Lr Sackville, NS B4C 2S6
(902) 864-4751
cleaners (Eco-Clean)

Enviro-Safe Products
1302 Gainsborough Dr.
Oakville, ON L6H 2H6
(416) 845-2598
cleaners & laundry products (Universal Soap)

Enviropro
P.O. Box 160
Alexandria, ON K0C 1A0
(613) 874-2301
cleaners; laundry products (Envirosoap); odour eliminators (Good Scent/ Bon Scent); natural dyes (Limona); cleaning mitt (Softy)

Frank T. Ross & Sons
P.O. Box 248
6550 Lawrence Ave. E.
West Hill, ON M1E 4R5
(416) 282-1107
cleaners & laundry products (Easy Clean, Nature Clean)

Grime Eater Products
1283 Matheson Blvd. E.
Mississauga, ON L4W 1R1
(416) 629-1212
cleaners; laundry products

Kirby & Wilson Manufacturing
925 Meyerside Dr.
Mississauga, ON L5T 1R8
(416) 670-1446
Cleaners (EnviroGREEN)

Laboratories Druide
202 – 2795 Bates Rd.
Montreal, PQ H3S 1B5
(514) 731-3815
cleaners & laundry products (Bio-Max)

Livos Canada Natural Structures & Supplies
P.O. Box 220, Station A
Fredericton, NB E3B 4Y9
(509) 366-3529
cleaners; furniture & floor polishes; leather seal; wax remover

Meter Pak
1A – 1011 Haultain Ct.
Mississauga, ON L4W 2K1
(416) 624-0366
cleaners

Mia Rose Products
C – 1374 Logan
Costa Mesa, CA 92629 USA
(714) 662-5465
air fresheners (Air Therapy); cleaners (Citri-Shine)

Mountain Fresh Products
P.O. Box 40516
Grand Junction, CO 81504 USA
(303) 434-8491
cleaners (Kleer, Kleen, Glass Mate); laundry products (Winter White)

Pets'n People
5312 Ironwood
Rancho Palos Verdes, CA 90274 USA
(213) 373-1559
cleaners; deodorizers

Prime Pacific
P.O. Box 87076
North Vancouver, BC V7L 4P6
(604) 929-7019
dishwasher & laundry products (Bio Dish)

Scandinavian Natural Health & Beauty Products
13 North Seventh St.
Perkasie, PA 18944 USA
(215) 453-2505
laundry products (Eco-Force)

Shaklee Canada
952 Century Dr.
Burlington, ON L7L 5P2
(416) 681-1422
cleaners; furniture polish; laundry products

The Soap Factory
141 Cushman Rd.
St. Catharines, ON L2M 6T2
(416) 682-1808
cleaners & laundry products (also Echo-Logic)

Teekah Environmental Products
96 Harbord St.
Toronto, ON M5S 1G6
(416) 962-3485
cleaners & laundry products

Van-l Products
562 John St.
Victoria, BC V8T 1T6
(604) 382-7342
cleaners & laundry products (Granny, V.I.P.)

Virginia's Soap
Group 60, Box 20, R.R. 1
Anola, MB R0E 0A0
(204) 866-3788
air fresheners, cleaners & pet care products (Next to Nature)

WHX Industries
1321 Sherbrooke St. West
Suite E-1
Montreal, PQ H3G 1J4
(514) 284-9512
cleaners (Citrikleen)

* * * * * * * * * * * * * * * * * * *

Cloth Diapers and Accessories

This category includes cloth diapers, diaper covers, diaper liners and aids for nursing mothers. Most companies provide catalogues of their products upon request. For diaper services, check the Yellow Pages in your phone book.

Absolutely Diapers
25 Ripley Ave.
Toronto, ON M6S 3P2
(416) 762-6650
diapers (Snap-To-Fit); diaper covers (R. Duck)

Altrim
450 Beaumont
Montreal, PQ H3N 1T7
(514) 273-8896
diapers, diaper covers & diaper liners (Baby Bliss, Bébé Calin)

Babykins Products
4 – 3531 Jacombs Rd.
Richmond, BC V6V 1Z8
(800) 665-2229
diapers

Baby Love Products
5015 – 46 St.
Camrose, AB T4V 3G3
(403) 672-1763
diapers; diaper liners; nursing aids

Born to Love
61 – 21 Potsdam Rd.
Downsview, ON M3N 1N3
(416) 663-7143
diapers (Every-Dry Liners); diaper covers (Down Unders); nursing pads (Nankins)

Bummis
C.P. 201, Station a
Montreal, PQ H3C 2S1
(514) 528-9438
diaper covers

The Crabapple Cotton Diaper Company
131 Mullen Dr.
Ajax, ON L1T 2A9
(416) 427-4931
diapers & diaper covers (Seedlings)

Diana Dolls Fashions
197 Barton Rd. E.
Stoney Creek, ON L8E 2K3
(416) 662-4281
diapers & nursing pads (Kooshies, Cushies, Diaper Plus)

Diaper Vision
3 – 275 Gordon St.
Fergus, ON N1M 2W3
(519) 843-5170
diapers

Eco-Bébé
20 Andras
Dollard-des-Ormeaux, PQ H9B 1R6
(514) 421-6990
diapers (Babe-Eez)

ECO ECO
1 – 3511 Hutchison
Montreal, PQ H2X 2G9
(514) 285-8881
unbleached diapers

Klas-Port
433 Connaught Ave.
North York, ON M2R 2M4
(416) 730-9023
diapers (Kom-Fort)

Lamasz
8001 Jane St.
Concord, ON L4K 2M7
(416) 660-4800
diapers

Manitou Designs
1460 Myron Dr.
Mississauga, ON L5E 2N6
(416) 274-8144
diapers (Cotton Tot's)

The Orchard Millworks
20 Orchard Mill Cres.
Kitchener, ON N2P 1T2
(519) 748-4117
diapers (Punkins)

Superfit Products
526 Rosedale Ave.
Winnipeg, MB R3L 1M3
(204) 284-1586
diaper covers & patterns

T's for Tots
25 Fairleigh Ave. S.
Hamilton, ON L8M 2K1
(416) 545-9989
diapers

* * * * * * * * * * * * * * * * * *

Composting Materials

Composting is a great way to reduce waste and provide your garden with rich fertilizer. This category includes composting bins, vermicomposting materials and related accessories.

Agri-Mart
9302 Denton Ave.
Hudson, FL 34667 USA
(813) 869-1212
composter starter (Comtrific)

Barry Bins
R.R. 3
Powassan, ON P0H 1Z0
(705) 724-2866
composters

Butler & Baird Lumber
P.O. Box 278
Aurora, ON L4G 3H4
(416) 727-3074
composters (Bio-Bin)

Canadian Original Vermicomposter
771 Coxwell Ave.
Toronto, ON M4C 3C6
(416) 469-2089
vermicomposting materials

Canagro Agricultural Products
P.O. Box 335
Elmira, ON N3B 2Z7
(519) 669-1586
compost accelorators

Composting Systems
P.O. Box 1507
Belleville, ON K8P 5J2
(613) 966-3727
composters (Bio-Keg)

Earth-Wise
R.R. 2
Roslin, ON K0K 2Y0
(613) 477-2693
composters

Eco Balance
84 Woodborough Rd.
Guelph, ON N1G 3K5
(519) 767-2644
composters

Environ-Mate
P.O. Box 272
West Hill, ON M1E 4R5
(416) 438-2548
compost aerators (Com-Post); compost bags (Air-Post, Com-Post)

Grow-Rich
8923 Chippewa Creek Rd.
Niagara Falls, ON L2E 6S5
(416) 357-6421
composters

The Natural Way
50 Bedford Rd.
Sutton West, ON L0E 1R0
(416) 722-9132
compost starter (Biplantol Plus)

Ritchers
P.O. Box 26
Goodwood, ON L0C 1A0
(416) 640-6677
compost activator

Rubbermaid
2562 Stanfield Rd.
Mississauga, ON L4Y 1S5
(416) 279-1010
compost accelerator, kitchen caddy & food digestor (Rubbermaid Green Cone)

T.S.E.
815 Center St.
North Manikato, MN 56001 USA
(507) 388-5424
composter plans (E-Z Composter)

Techstar Plastics
15400 Old Simcoe Rd.
Port Perry, ON L9L 1L8
(800) 263-7943
composters (Bardmatic)

Teekah Environmental Products
96 Harbord St.
Toronto, ON M5S 1G6
(416) 962-3485
composters

* * * * * * * * * * * * * * * * * * *

Energy-Saving Devices

Energy-saving devices include efficient lighting, heating, windows, and doors. Also included in this category are programmable devices such as thermostats and light timers. As with any purchase, before you buy check the warranty of the product. Look for those with extended warranties and with exceptional performance records. Check consumer magazines for comparison tests. Simply, buy products that last.

Bestek Micro Devices & Systems
19 – 2485 Lancaster Rd.
Ottawa, ON K1B 5L1
(613) 523-5211
fluorescent lighting & solar power products (Solbrite)

Darentek
5 Frederick Place
Ottawa, ON K1S 3G1
(613) 234-9041
solar power products

E.L. Foust Company
P.O. Box 105
Elmhurst, IL 60126 USA
(800) 225-9549
foil vapour barrier (Dennyfoil); aluminum-foil tape (Polyken); steam humidifier (Skuttle)

Humico
256 South Robertson
Beverly Hills, CA 90211 USA
(213) 856-0821
refrigerator (Food Saver)

OSRAM
6185 Tomken Rd.
Mississauga, ON L5T 1X6
(416) 670-1996
fluorescent lighting (Dulux E1)

Peka Rollshutters
1246 3rd Ave. N.
Lethbridge, AB T1K 4H3
(403) 329-4597
insulated rollshutters

Pella-Hunt
1040 Wilton Grove Rd.
London, ON N6A 4C2
(519) 686-3100
door & windows (Hunt, Pella)

* * * * * * * * * * * * * * * * * * *

Home Building and Renovation

Look for paint, caulking compounds, adhesives, stains and finishes, etc. that are made of natural ingredients (rather than synthetic) and that do not contain questionably safe chemicals. To determine if the product does contain potentially hazardous ingredients, read the label. Look for warnings such as poisonous, flammable, or corrosive. When possible, look for products (such as wood) that contain recycled content.

Canadian Old Fashioned Milk Paint Co.
163 Queen St. E.
Toronto, ON M5A 1S1
(416) 364-1393
paints

Deft
17451 Von Karman Ave.
Irvine, CA 92714 USA
(714) 474-0400
stains

Frank T. Ross & Sons (1962) Ltd.
P.O. Box 248
6550 Lawrence Ave. E.
West Hill, ON M1E 4R5
(416) 282-1107
adhesives (Weldbond)

Nitty Gritty Reproductions
163 Queen St. E.
Toronto, ON M5A 1S1
(416) 364-1393
formaldehyde-free milk-painted furniture

Prosumex
820 Ellingham
Pointe Claire, PQ H9R 3S4
(514) 694-1485
cellulose wood fibres (Privex dB-2000)

The Natural Stain Co. of Canada Inc.
P.O. Box 244
Cochrane T0L 0W0
Woods: The Natural Choice Wood Conditioner and Stain

Teekah Environmental Products
96 Harbord St.
Toronto, ON M5S 1G6
(416) 962-3485
paints (Auro)

* * * * * * * * * * * * * * * * * *

Lawn and Garden

When buying products for your lawn and garden, read the labels. Look for products made of natural ingredients and that are plant-based rather than petroleum-based. Some of these products may contain pyrethrins, which have been proven moderately toxic to humans. Also check the label for warnings such as "poisonous," "flammable," or "corrosive." Buy in large sizes when possible to minimize packaging and look for packaging that is refillable, contains recycled content, or is recyclable.

A-GRO-ELITE
308 – 265 S. Federal Hiway
Deerfield Beach, FL 33441 USA
fertilizers (also Maestro-Gro)

Aerokure
P.O. Box 22
Sherbrooke, PQ J1H 5H5
(819) 821-2238
pest control devices (Aeroxon, Hortikure)

Agri-Mart
9302 Denton Ave.
Hudson, FL 34667 USA
(813) 869-1212
fertilizers (Agrimart Plant Shield, AM, Flourish, Micro Plex Plus, Pot Shot, Soil Treatment, Thrive); pesticides (Newmout)

Beneficial Insectary
14751 Oak Run Rd.
Oak Run, CA 96069 USA
(916) 472-3715
beneficial insects

Benmax
6721 est, rue Beaubien
Montreal, PQ H1M 3B2
(514) 259-2561
pest control devices

Canagro Agricultural Products
P.O. Box 335
Elmira, ON N3B 2Z7
(519) 669-1586
fertilizers (Calcitic, 1 Kare, Nature's Fertilizer)

Chemfree Environment
16763 Hymus Blvd.
Kirkland, PQ H9H 3L4
(514) 630-4400
insecticides (Insectigone)

Dynamic Essentials
3 Waterloo St.
New Hamburg, ON N0B 2G0
(519) 662-2520
watering system

EMJ Environmental Products
R.R. 6, Highway 24 South
Guelph, ON N1H 6J3
(519) 837-2444
pesticides (Brown Bag)

EcoGenesis
16 Jedburgh Rd.
Toronto, ON M5M 3J6
(416) 489-9031
fertilizer (Restore); seeds

Green Cross
6860 Century Ave.
Mississauga, ON L5N 2W5
(416) 821-4430
insecticides & pest control devices

Grow-Rich
8923 Chippewa Creek Rd.
Niagara Falls, ON L2E 6S5
(416) 357-6421
compost, fertilizers & soil (Grow-Rich)

Nutrite Fertilizers
23 Union St., Box 160
Elmira, ON N3B 2Z6
(519) 669-5401
fertilizers

Santa Barbara Greenhouses
J – 1115 Acaso Ave.
Camarillo, CA 93010 USA
(805) 482-3765
greenhouse & mist systems

****

Paper and Related Products

Whenever possible, choose paper products containing recycled rather than virgin pulp. Look for the symbol indicating that the product contains recycled material. However, be aware that a product containing only 5% post-consumer waste carries the same symbol as a product that contains 100% post-consumer waste. When you buy a product, read the label and choose the one that has the highest recycled content. Also look for paper products that are dioxin-free (unbleached or oxygen-bleached), undyed, and unscented.

The Almost Perfect Packaging Company
204 – 35 Macdonell St.
Guelph, ON N1H 2Z4
(519) 763-1490
stationery

Anthony's Originals
P.O. Box 8336-GP
Natik, MA 01760 USA
(508) 655-8937
reusable bag patterns (Freeby-Bags); reusable envelope patterns (Envy-Lopes)

Atlantic Packaging Products
111 Progress Rd.
Scarborough, ON M1P 2Y9
(416) 298-5427
stationery

Brush Dance
218 Cleveland Ct.
Mill Valley, CA 94941 USA
(415) 389-6228
stationery; wrapping paper

Courseware Solutions
404 – 100 Lombard St.
Toronto, ON M5C 1M3
(416) 863-6116
office paper (Tree Saver)

Hai-Tech Instruments
P.O. Box 334, Station NDG
Montreal, PQ H4A 3P6
(514) 485-4570
stationery (Acorn Designs)

Innova Envelope
56 Steelcase Rd. W.
Markham, ON L3R 1B2
(416) 475-6181
envelopes (Heritage)

LaserNetworks
1 – 785 Pacific Rd.
Oakville, ON L6L 6M3
(416) 847-7823
toner cartridges

Teekah Environmental Products
96 Harbord St.
Toronto, ON M5S 1G6
(416) 962-3485
envelopes, stationery and office paper

****

Personal Care

This category includes cosmetics, body care, baby care, and feminine hygiene products (including reusable menstrual pads). Buy in the largest size available to minimize packaging. Look for products that are made of natural ingredients (rather than synthetic chemicals), readily biodegradable, vegetable oil-based (rather than mineral oil-based), unscented, free of dyes or colouring, talc-tree (body powders), cruelty-free (not tested on animals), and packaged in reusable, recycled, or recyclable packaging.

Abkit
1160 Park Ave.
New York, NY 10128 USA
(212) 860-8358
body, facial & hair care (CamoCare)

Alldec Trading Co.
110 – 19329 Enterprise Way
Surrey, BC V3S 6J8
(604) 534-3688
barrier cream (pr88)

Anthony's Originals
P.O. Box 8336-GP
Natik, MA 01760 USA
(508) 655-8937
reusable shaving cream foamer/dispenser (No-Fuss-Foam)

Aura Cacia
P.O. Box 399
Weaverville, CA 96093 USA
(916) 623-3301
body care; perfumes

Autumn-Harp
28 Rockydale Rd.
Bristol, VT 05443 USA
(812) 453-4807
baby care; body & facial care (also Lipsense, Ultracare, Un-Petroleum Jelly); hair care

Beauty and the Beach
2277 Queen St. E.
Toronto, ON M4E 1G5
(416) 698-2944
baby, facial & hair care (Bushwacky, H.E.A.R.T.); cosmetics

The Body Shop
15 Prince Andrew Place
Don Mills, ON M3C 2H2
(416) 441-3202
body, facial & hair care; cosmetics; perfumes

Caeran
25 Penny Lane
Brantford, ON N3R 5Y5
(519) 751-0513
body, facial & hair care

Clientele
5207 NW 163 St.
Miami, FL 33014 USA
(305) 624-6665
body & facial care

Crabtree & Evelyn
1010 Adelaide St. S.
London, ON N6E 1R6
(519) 685-1112
baby, body, facial & hair care

Cruickshank's
1015 Mount Pleasant Rd.
Toronto, ON M4P 2M1
(416) 488-8292
body care (also Janet Carter); insect repellant (Janet Carter)

DTR Dermal Therapy Research
2 – 30 Pacific Ct.
London, ON N5V 3K4
(519) 453-5190
body & facial care

EMJ Environmental Products
R.R. 6, Highway 24 S.
Guelph, ON N1H 6J3
(519) 837-2444
body care (Simply Clean)

Emme Imports
67 Poplar Heights Dr.
Islington, ON M9A 4Z3
(416) 240-0647
body, facial & hair care (IDI Pharmaceuticals)

Enviro Safe Products
1302 Gainsborough Dr.
Oakville, ON L6H 2H6
(416) 845-2598
body care (Liquid Hand & Body Soap, Royal Alpine Bar Soap); hair care (Herbal Shampoo)

Frank T. Ross & Sons
P.O. Box 248
6550 Lawrence Ave. E.
West Hill, ON M1E 4R5
(416) 282-1107
hair care (Nature Clean)

Grime Eater Products
1283 Matheson Blvd. E.
Mississauga, ON L4W 1R1
(416) 629-1212
body & facial care

Heritage Healing Herbs
P.O. Box 78021
1547 Merivale Rd.
Nepean, ON K2G 5W2
(615) 727-5783
baby, body & hair care (also Home Health); feminine hygiene (Kress & Owen)

I & M Natural Skincare
79A Harbord St.
Toronto, ON M5S 1G4
(416) 975-4465
body, facial & hair care

Indian Creek Naturals
P.O. Box 63
Selma, OR 97538 USA
(503) 592-2616
body & facial care

International Cosmeticare
3 – 12240 Horseshoe Way
Richmond, BC V7A 4X9
(604) 277-6687
hair care (Herbal Glo)

The Keeper
P.O. Box 20023
Cincinnati, OH 45220 USA
(513) 221-1464
feminine hygiene products

Kirby & Wilson Manufacturing
925 Meyerside Dr.
Mississauga, ON L5T 1R8
(416) 670-1446
body & hair care (EnviroGREEN)

Laboratories Druide
202 – 2795 Bates Rd.
Montreal, PQ H3S 1B5
(514) 731-3815
baby, body & hair care (Druide)

Lange' Laboratories
21093 Forbes Ave.
Hayward, CA 94545 USA
(415) 785-6570
hair & facial care

Mountain Fresh Product
P.O. Box 40516
Grand Junction, CO 81504 USA
(303) 434-8491
baby care (Baby Massage); body & hair care (Aloe Gold, Golden Lotus, Vegelatum)

Mira Linder Spa In The City
108 Avenue Rd
Toronto, ON M5R 2H3
(416) 961-6900
body & facial care; cosmetics

No Common Scents
220 Xenia Ave., Kings Yard
Yellow Springs, OH 45387 USA
(513) 767-4261
insect repellent

Prima Vera Products
22 Howland Ave.
Toronto, ON M5R 3B3
(416) 536-5011
baby, body, facial & hair care

Rainbow Research Corp.
170 Wilbur Place
Bohemia, NY 11716 USA
(516) 589-5563
body, facial & hair care (Rainbow, Stony Brook Botanicals)

Revlon
2501 Stanfield Rd.
Mississauga, ON L4Y 1R9
(416) 276-4500
hair care (Revlon Milk Plus 6)

Seasons Cosmetics
5650 Tomken Rd., Unit 7
Mississauga, ON L4W 4P1
(416) 890-2255
cosmetics; facial care; perfumes

The Soap Factory
141 Cushman Rd.
St. Catharines, ON L2M 6T2
(416) 682-1808
body & hair care (also Echo-Logic)

Soapberry Shop
12 – 50 Galaxy Blvd.
Rexdale, ON M9W 4Y5
(416) 674-0248
body, facial & hair care; cosmetics

T's for Tots
25 Fairleigh Ave. S.
Hamilton, ON L8M 2K1
(416) 545-9989
adult sanitary products; feminine hygiene

Truso
33 Sheperd Rd.
Oakville, ON L6K 2G6
(416) 844-8144
hair care (Enviro-Perm)

Virginia's Soap
Group 60, Box 20, R.R. 1
Anola, MB R0E 0A0
(204) 866-3788
body & facial care (also Next to Nature)

Visage Cosmetics
801 Eglinton Ave. W.
Toronto, ON M5N 1E3
(416) 789-7191
cosmetics & facial care (Caryl Baker Visage)

Walnut Acres Organic Farms
Walnut Acres Rd.
Penns Creek, PA 17862 USA
(717) 837-0601
baby, body & hair care (also Weleda)

Women's Choice
P.O. Box 245
Gabriola, BC V0R 1X0
(604) 247-8433
feminine hygiene

* * * * * * * * * * * * * * * * * *

Textiles

This category includes clothing, dyes, cotton bags, carpets, drapes, and upholstery products. Although the choice is not always clear, as a general rule choose natural fibres over synthetic ones. Look for fabrics that are unbleached and undyed whenever possible. In dyed fabrics, look for vegetable-based (rather than petroleum-based) colourings. Choose organically grown cotton.

The Almost Perfect Packaging Company
204 – 35 Macdonell St.
Guelph, ON N1H 2Z4
(519) 763-1490
clothing; tote bags; towels

Anthony's Originals
P.O. Box 8336-GP
Natik, MA 01760 USA
(508) 655-8937
tote bag patterns (Freeby Bags)

Coffee Sock Co.
P.O. Box 10023
Eugene, OR 97440 USA
(503) 344-7698
cloth coffee filters

ECO ECO
1 – 3511 Hutchison
Montreal, PQ H2X 2G9
(514) 285-8881
tote bags

Earth-Wise
R.R. 2
Roslin, ON K0K 2Y0
(613) 477-2693
cloth coffee socks; clothing; tote bags

Enviro Wear
5 – 2364Haines Rd.
Mississauga, ON L4Y 1Y6
(416) 566-8181
clothing; tote bags

H-T Ecologic Product
5600 Place Beaminster
Montreal, PQ H3W 2M3
(514) 739-5889
clothing

Industrial Wear
P.O. Box 1349
Studio City, CA 91614 USA
(818) 508-7324
tote bags

Lamasz
8001 Jane St.
Concord, ON L4K 2M7
(416) 660-4800
clothing (Snow Wear); tote bags

Nikki
P.O. Box 6730, Station J
Ottawa, ON K2A 3Z4
(613) 596-1816
tote bags

Pacific Wellspring
P.O. Box 27037
Victoria, BC V9B 5S4
(604) 743-9962
tote bags

Rescue Earth
1525 Hillsmont Dr.
El Cajon, CA 92020 USA
(619) 588-7700
tote bags

Teekah Environmental Products
96 Harbord St.
Toronto, ON M5S 1G6
(416) 962-3485
cloths; tote bags

Treekeepers
518 – 249 S. Highway 101
Solana Beach, CA 92075 USA
(619) 481-6403
tote bags

* * * * * * * * * * * * * * * * * * *

Water-Saving Devices

These include low-flow showerheads, faucets, and toilets. As with any purchase, befor you buy check the warranty of the product. Look for those with extended warranties and with exceptional performance records. Check consumer magazines for comparison tests. Simply, buy products that last.

Aerokure
P.O. Box 22
Sherbrooke, PQ J1H 5H5
(819) 821-2238
compost toilets (Septi-Kure)

BMT Environmental Technologies
P.O. Box 498
Toronto, ON M4G 4E1
(416) 925-1668
aerators; faucets; showerheads; toilets; toilet dams

Composting Toilet Systems
1211 Bergen Rd.
Newport, WA 99156-9608 USA
(509) 447-3708
compost toilets

Dynamic Essentials
3 Waterloo St.
New Hamburg, ON N0B 2G0
(519) 662-2520
faucets (Spraysys); showerheads (Showersys); watering systems (Terra-Sys)

Enviropro
P.O. Box 160
Alexandria, ON K0C 1A0
(613) 874-2301
showerheads & toilet dams (Maximizer)

Newfound Trading
206 – 11 Morris Dr.
Dartmouth, NS B3B 1M2
(902) 468-7100
automatic waterstop

Sancor Industries
140 – 30 Milner
Scarborough, ON M1S 3R3
(416) 299-4818
waterless toilets (Envirolet)

Sanitation Equipment
35 Citron Ct.
Concord, ON L4K 2S7
(416) 738-0055
toilets (Flush-O-Matic #707)

Sun-Mar
C2 – 5035 North Service Rd.
Burlington, ON L7L 5V2
(416) 332-1314
compost toilets

Transcontinential Energy Saving Products
7 – 4179 Harvester Rd.
Burlington, ON L7L 5M4
(416) 639-0937
showerheads (Shower Saver)

* * * * * * * * * * * * * * * * * * *

Miscellaneous

Atlantic Packaging Company
204 – 35 Macdonell St.
Guelph, ON N1H 2Z4
(519) 763-1490
giftbaskets; stationery; tote mugs

Authentic Euthenics
125 – 4951 Clairmont Square
San Diego, CA 92117 USA
(619) 270-7056
can compactor (The Crusher, Easy Crush); newspaper bundler (Stack-N-Tie); recycling bins (Easy Sort)

Better World Investment Corp.
203 – 276 Carlaw Ave.
Toronto, ON M4M 3L1
(416) 465-4431
air pressure-based plunger (Pango)

Bioman Products
4 – 400 Matheson Blvd.
Mississauga, ON L4Z 1N8
(416) 890-2555
test kits for food analysis

Conros
1190 Birchmount Rd.
Scarborough, ON M1P 2B8
(416) 751-4343
firelogs (Northland Clean-burn)

Earth-Wise
R.R. 2
Roslin, ON K0K 2Y0
(613) 477-2693
newspaper bundler; plastic-bag holder

The Kit Company
128 Walmer Rd.
Toronto, ON M5R 2X9
(416) 922-3654
The Green House Game

LifeFilter
829 Richmond St.
London, ON N6A 3H7
(519) 660-6613
permanent coffee filters

Livos Canada Natural Structures & Supplies
P.O. Box 220, Station A
Fredericton, NB E3B 4Y9
(506) 366-3529
art materials

Rain Forest Crunch
R.D. 2, P.O. Box 1950
Mount Peliere, VT 05602 USA
(802) 229-1840
candy made from rain forest products

Recycle Resources
P.O. Box 7295
Salem, OR 97303 USA
(503) 585-6741
recycling bin

Safe Co.
368 Hillside Ave.
Needham, MA 02194 USA
(617) 444-7778
radiation monitor

Environmental Outlets

Listed below are stores and mail-order outlets devoted to carrying environment-friendly products. Many of the products listed in the preceding sections are available through these outlets. Natural or health food stores often carry many of these products as well, but they are too numerous to list here — check the Yellow Pages of your phone book. Product catalogues are usually available and often free. Just ask!

Alternatives Market
579 Kerr St.
Oakville, ON L6K 3E1
(416) 844-2375

All Things Wise and Wonderful
P.O. Box 267, Pierrefonds Station
Pierrefonds, PQ H9H 4K9
(514) 683-8407

Back to Basics Environment Store
1743 Carling Ave.
Ottawa, ON K2A 1C8
(613) 722-6554

B.C. Solar
P.O. Box 317
Horsefly, BC V0L 1L0
(604) 620-3510

Dodie's Biodegradables
17 Queen Anne Rd.
Toronto, ON M8X 1T1
(416) 233-7112

Earth Care Paper Inc.
P.O. Box 14140
Madison, WI 53714-0140 USA
(608) 277-2900

Earth-Wise
R.R. 2
Roslin, ON K0K 2Y0
(613) 477-2693

Earthright
P.O. Box 400
Horsefly, BC V0L 1L0
(604) 620-3510

Earthwise Books
216 Bank St.
Ottawa, ON K2P 1X1
(613) 238-8363

Ecco Bella
602 – 6 Provost Square
Caldwell, NJ 07006 USA
(201) 226-5799

Harvest Collective
877 Westminster
Winnipeg, MB R3G 1B3
(204) 772-4359

Healthy Solutions
35 Clinton St.
Toronto, ON M6J 2N9
(416) 537-8986

The Natural Order
P.O. Box 850, Station P
Toronto, ON M5S 2Z2
(416) 975-1682

The Paper Source
Fallbrook, ON K0G 1A0
(613) 267-7191

Teekah Environmental Products
96 Harbord St.
Toronto, ON M5S 1G6
(416) 962-3485

* * * * * * * * * * * * * * * * * *

Services

The following services include asbestos abatement & removal, building renovation, home/office cleaning, laboratory testing (air, soil, water, etc.), recycling centres, radon testing and/or mitigation.

Diaper services are too numerous to list and can be found in the Yellow Pages of your phone book.

A.C.C. Encon
108 Gallery Square
Montreal, PQ H3C 3R3
(514) 937-6151
asbestos abatement & removal

Abbatec
2615 – 103A Blackwell Ave.
Ottawa, ON K1B 4E4
(613) 741-3372
recycling centre: photography chemicals, plastic, silver

Action Print & Graphics
318 Richmand St. W.
Toronto, ON M5V 1X2
(416) 596-8015
printing

Aerovac Services
358 E. 12th St.
North Vancouver, BC V7L 2J9
(604) 988-0615
asbestos abatement & removal; home/ office cleaning

Alberta Recoveries & Rentals
General Delivery
Medicine Hat, AB T1A 7E4
(403) 527-7003
recycling centre: newsprint/cardboard, plastic (raw material)

All Ports Environmental
284 Wilson St. E.
Hamilton, ON L8L 1S3
(416) 527-7100
asbestos abatement & removal;
laboratory testing

Asbestez Distributor
3427 – 12 St. N.E.
Calgary, AB T2E 6S6
(403) 250-3169
asbestos abatement & removal

Atlantic Packaging Products
111 Progress Ave.
Scarborough, ON M1P 2Y9
(416) 298-5427
recycling centre: newsprint/cardboard

Bioman Products
4 – 400 Matheson Blvd.
Mississauga, ON L4Z 1N8
(416) 890-2555
laboratory testing

Canadian Asbestos Services
102 – 190 Stafford Rd. W.
Nepean, ON K2H 9G3
(613) 596-9665
asbestos abatement & removal

Canadian Eagle Recyclers
16 Melanie Dr.
Brampton, ON L6T 4K9
(416) 292-2209
recycling centre: asphalt, concrete, doors,
lumber, sinks, steel beams, wood waste

Canadian Fibre
3971 Boundary Rd.
Richmond, BC V6V 1T8
(614) 524-4627
recycling centre: aluminum & other
metal, fine paper, glass, newsprint/
cardboard, plastic

Consolidated Environment
415 Dawson Rd.
Winnipeg, MB R2J 0S8
(204) 237-3310
laboratory testing; recycling centre: oil

Da-Lee Dust Control
P.O. Box 14
350 Jones Rd.
Fruitland, ON L0R 1L0
(416) 643-1135
recycling centre: oil

Delta Recycling Society
8828 River Rd.
Delta, BC V4G 1B5
(604) 946-9828
recycling centre: aluminum & other
metal, fine paper, glass, newsprint/
cardboard

Fleck Contracting
204 – 8449 Main St.
Vancouver, BC V5X 3M3
(604) 324-2444
asbestos abatement & removal; building
renovation

Florence Paper Company
2475 Sheffield Rd.
Ottawa, ON K2B 3V6
(613) 745-9437
recycling centre: fine paper, newsprint/
cardboard

For Earth's Sake
80 Baker St.
Guelph, ON N1H 4G1
(519) 837-3242
building renovation; home/office
cleaning; laboratory testing; radon
testing and/or mitigation; recycling
centre: aluminum & plastic

The Garage of Kane
321 Winona Dr.
Toronto, ON M6C 3T2
(416) 975-3982
Freon recycling

Genor Services
434 Henry St.
Brantford, ON N3T 5W5
(519) 756-5264
recycling centre: aluminum & other
metals, fine paper, newsprint/
cardboard, plastics

Go-For Used Oil
625 MacDonald St.
Regina, SK S4N 4X1
(306) 721-6755
oil reclaimer

Grandma's Company
P.O. Box 4
Nestleton, ON L0B 1L0
(416) 986-0689
home/office cleaning

Green City Design & Construction
12 Newmarket Ave.
Toronto, ON M4C 1V7
(416) 691-2477
building renovation

Griffin Laboratories
1875 Spall Rd.
Kelowna, BC V1Y 4R2
(604) 861-3234
laboratory testing; radon testing and/or
mitigation

I.G. Machine & Fibres
87 Orenda Rd.
Brampton, ON L6W 1V7
(416) 457-0745
recycling centre: fine paper, newsprint/
cardboard

MacDonald's Industrial Services
R.R. 5
New Glasgow, NS B2H 5I8
(902) 992-3007
asbestos abatement & removal

New West Gypsum
20321 – 80th Ave.
Langley, BC V3A 4P7
(604) 888-2282
recycling centre: drywall board

New Trend Oil
788 Winger Rd.
Williams Lake, BC V2G 3S4
(604) 392-2724
recycling centre: oil

Newfoundland Envirotech
P.O. Box 13192
St John's, NF A1B 4A4
(709) 753-6601
asbestos abatement & removal; radon
testing and/or mitigation

Nilfisk
200 – 7 Connie Crescent
Concord, ON L4K 1M1
(416) 669-6003
asbestos abatement & removal; building
renovation; home/office cleaning

Norwest Labs
203 – 20771 Langley Bypass
Langley, BC V3A 5E8
(604) 530-4344
laboratory testing

North West Consulting and Testing
P.O. Box 336, Station E
Victoria, BC V8W 2N2
(604) 384-9695
asbestos abatement & removal

Paper Chase Recycling
12155 William Short Rd.
Edmonton, AB T5B 2E1
(403) 477-9391
recycling centre: fine paper, newsprint/ cardboard

Reclaim Enterprises
P.O. Box 8347
Dundas, ON L9H 6M1
(416) 627-9170
asbestos abatement & removal; recycling centre (mobile equipment): aluminum & other metal, automobiles, newsprint/ cardboard

Recyclage P.F.
C.P. 262, 10 rue Du Ruisseau
Port Cartier, PQ G5B 2G8
(418) 766-6581
recycling centre: aluminum & other metal, automobiles, fine paper, newsprint/cardboard, plastics

Ridge Meadows Recycling Society
P.O. Box 283
Maple Ridge, BC V2X 7G2
(604) 463-5545
recycling centre: aluminum & other metal, fine paper, glass, newsprint/ cardboard, oil

Safety Express
12 – 2140 Winston Park Dr.
Oakville, ON L6H 5V5
(416) 829-3777
asbestos abatement and removal

Secural Datashred
8 – 105 Riviera Dr.
Markham, ON L3R 5J7
(416) 940-3282
recycling centre: aluminum, fine paper, newsprint/cardboard

Tiger-Vac
11600, 6e Ave. R.D.P.
Montreal, PQ H1E 1S1
(514) 643-1525
asbestos abatement & removal; home/ office cleaning; radon testing

Tri-Tech Recycling
505 Dotzert Ct.
Waterloo, ON N2L 6A7
(416) 747-2226
recycling centre: aluminum & other metal, fine paper, glass, newsprint/ cardboard, plastic

Wearmouth Waste-Tech
P.O. Box 1, Site 1, R.R. 5
Calgary, AB T2P 2G6
(403) 236-2202
asbestos abatement & removal

Willpax
1415 Whyte Ave.
Winnipeg, MB R3E 1V7
(204) 772-5203
recycling centre: fine paper, polyethylene

Travel Agencies

In 1989, Canadian wildlife and naturalist groups were asked to name a few tour operators that had established a particularly good reputation among the conservation community. The following firms were recommended, but there are thousands of tour operators offering all manner of services. For more information, contact your provincial or federal tourism department.

Athabasca Trail Trips
Box 6117
Hinton, AB T7V 1X5
(403) 865-7549
horse-assisted hiking and camping (horses used to carry gear only)

The Adventure Centre
17 Hayden St.
Toronto, ON M4Y 2P2
(416) 922-7584
camping, cycling tours, trekking, overlanding

Black Feather
1341 Wellington St.
Ottawa, ON K1Y 3B8
(613) 722-9717
canoeing, hiking, kayaking, whitewater rafting

Canadian Himalayan Expeditions
721 Bloor St. W.
Toronto, ON M5G 1L5
(416) 537-2000
adventure travel: trekking, river rafting, jeep safaris

Canadian Nature Tours
355 Lesmill Rd.
Don Mills, ON M3B 2W8
(416) 444-8419
natural history tours: birdwatching, canoeing, cycling, scientific research expeditions, whalewatching

Canadian Wilderness Trips
171 College St.
Toronto, ON M5T 1P7
(416) 977-3703
birdwatching, camping, canoeing, hiking, outdoor school

East Wind Arctic Tours and Outfitters
Box 2728
Yellowknife, NWT X1A 2R1
(403) 873-2170
archaeological tours, backpacking, birdwatching, boating, hiking, photography tours, wildlife tours

Eco Summer Expeditions
1516 – AG Duranleau St.
Granville Island, Vancouver, BC V6H 3S4
(604) 669-7741
nature trips, photography tours, sailing, sea kayaking, whalewatching

Mingan Islands Cetacean Study
285 Green St.
St-Lambert, PQ J4P 1T3
(514) 465-9176 (winter)
(418) 949-2845 (summer)
trips to watch and study whales, dolphins, and other marine mammals

Ocean Search Tours
P.O. Box 129
Grand Manan, NB E0G 2M0
(506) 662-8144
birdwatching, sailing, whalewatching

Passages Exotic Expeditions
296 Queen St. W.
Toronto, ON M5V 2A1
(416) 593-0942
outdoor adventure holidays

Sobek Canada
159 Main St.
Unionville, ON L3R 2G8
(416) 479-2600
adventure holidays, kayaking, rafting, research holidays, trekking

Worldwide Adventures
747 – 920 Yonge St.
Toronto, ON M4W 3C7
(416) 963-9163
adventure holidays, birdwatching, gorilla-watching, rafting, sailing, trekking, whalewatching

****

Groups for More Information

Canadian Coalition on Acid Rain
401 – 112 St. Clair Ave. W.
Toronto, ON M4V 2Y3
(413) 968-2135

Canadian Environmental Network
P.O. Box 1289, Station B
Ottawa, ON K1P 5R3
(613) 563-2078
(inquire about provincial networks)

Canadian Federation of Humane Societies
102 – 30 Concourse Gate
Nepean, ON K2E 7V7
(613) 224-8072

Canadian Nature Federation
453 Sussex Dr.
Ottawa, ON K1N 6Z4
(613) 238-6154
(inquire about provincial and local naturalist groups)

Canadian Organic Growers
Box 6408, Station J
Ottawa, ON K2A 3Y6
(613) 259-2967

Canadian Parks and Wilderness Society
1150 – 160 Bloor St. E.
Toronto, ON M4W 1B9
(416) 972-0868
(inquire about provincial and local groups)

Canadian Wildlife Federation
1673 Carling Ave.
Ottawa, ON K2A 1C4
(613) 725-2191

Children's Rainforest
P.O. Box 936
Leniston, ME 04240 USA

Energy Probe
225 Brunswick Ave.
Toronto, ON M5S 2M6
(416) 978-7014

Environmental Youth Alliance
P.O. Box 29031, 1996 W. Broadway St.
Vancouver, BC V6J 5C2
(604) 737-2258

Environmentally Sound Packaging Coalition
2150 Maple St.
Vancouver, BC V6J 3T3
(604) 736-3644

Friends of the Earth
701 – 251 Laurier Ave. W.
Ottawa, ON K1P 5J6
(613) 230-3352

Greenpeace
185 Spadina Avenue, Sixth Floor
Toronto, ON M5T 2C6
(416) 345-8408

Pollution Probe
12 Madison Ave.
Toronto, ON M5R 2S1
(416) 926-1907

Probe International
225 Brunswick Ave.
Toronto, ON M5S 2M6
(416) 978-7014

Sierra Club of Eastern Canada
2316 Queen St. E.
Toronto, ON M4E 1G8
(416) 698-8446

Sierra Club of Western Canada
314 – 620 View St.
Victoria, BC V8W 3T8
(604) 386-5255

Société pour vaincre la pollution
CP 65, Succursale Place d'Armes
Montreal, PQ H2Y 3E9
(514) 844-5477

Solar Energy Society
3 – 15 York St.
Ottawa, ON K1N 5S7
(613) 236-4594

Temagami Wilderness Society
307 – 19 Mercer St.
Toronto, ON M5V 1H2
(416) 599-0152

Western Canada Wilderness Committee
20 Water St.
Vancouver, BC V6B 1A4
(604) 683-8220

World Wildlife Fund (Canada)
90 Eglinton Ave. E., Suite 504
Toronto, ON M4P 2N7
(416) 489-8800

World Society for the Protection of Animals
P.O. Box 15
55 University Ave., Suite 902
Toronto, ON M5J 2H7
(416) 369-0044

****

Provincial Naturalist Groups

Newfoundland Natural History Society
P.O. Box 1013
St. John's, NF A1C 5M3

Nova Scotia Bird Society
c/o The Nova Scotia Museum
1747 Summer St.
Halifax, NS B3H 3A6
(902) 429-4610

New Brunswick Federation of Naturalists
277 Douglas Ave.
Saint John, NB E2K 1E5

Natural History Society of Prince Edward Island
53 Fitzroy St.
Charlottetown, PE C1A 1R4

Union québécoise pour la conservation de la nature
160 – 76 St. E., 2nd floor
Charlesbourg, PQ G1H 4R3
(418) 628-9600

Province of Quebec Society for the Protection of Birds
240 – 36th Ave.
Lachine, PQ H8T 2A3
(514) 637-2141

Federation of Ontario Naturalists
355 Lesmill Rd.
Don Mills, ON M3B 2W8
(416) 444-8419

Manitoba Naturalists Society
302 – 128 James Ave.
Winnipeg, MB R3B 0N8

Saskatchewan Natural History Society
P.O. Box 4348
Regina, SK S4P 3W6
(306) 665-1915

Federation of Alberta Naturalists
P.O. Box 1472
Edmonton, AB T5J 2N5
(403) 453-8629

Ecology North
P.O. Box 2888
Yellowknife, NT X1A 3S9
(403) 873-6019

Federation of British Columbia Naturalists
321 – 1367 W. Broadway
Vancouver, BC V6H 4A9
(604) 737-3057

****

Environmental Networks

Atlantic Environmental Network
3115 Veith St., 3rd Floor
Halifax, NS B3K 3G9
(902) 454-2139

Réseau québécois des groupes écologistes
CP 1480, Succursale Place d'Armes
Montreal, PQ H2Y 3K8
(514) 982-9444

Ontario Environment Network
201C – 2 Quebec St.
Guelph, ON N1H 2T3
(519) 837-2565

Manitoba Eco-Network
P.O. Box 3125
Winnipeg, MB R3C 4E6
(204) 956-1468

Saskatchewan Eco-Network
P.O. Box 1372
Saskatoon, SK S7K 0G4
(306) 665-1915

Alberta Environment Network
10511 Saskatchewan Dr.
Edmonton, AB T6E 4S1
(403) 465-0872

Northern Environmental Network
P.O. Box 4163
Whitehorse, YT Y1A 3S9
(403) 668-5687

British Columbia Environmental Network
2150 Maple St.
Vancouver, BC V6J 3T3
(604) 733-2400

****

Recycling Organizations

Ecology Action Centre
3115 Veith St., 3rd Floor
Halifax, NS B3K 3G9
(902) 454-7828

Fonds québécois de récuperation
500 – 407 Saint Laurent Blvd.
Montreal, PQ H2Y 2Y5
(514) 874-3701

Recycling Council of Ontario
504 – 489 College St.
Toronto, ON M6G 1A5
1-800-263-2849

Recycling Council of Manitoba
412 McDermot Ave.
Winnipeg, MB R3A 0A9
(204) 942-7781

Recycling Council of British Columbia
2150 Maple St.
Vancouver, BC V6J 3T3
(604) 731-7222

Federal Government

Department of Environment
Terrasses de la Chaudière
10, rue Wellington
Hull, PQ K1A 0H3
(819) 997-2800

External Affairs and International Trade Canada
Lester B. Pearson Bldg.,
125 Sussex Dr.,
Ottawa, ON K1A 0G2
(613) 996-9134

Energy, Mines and Resources Canada
580 Booth St.
Ottawa, ON K1A 0E4
(613) 995-3065

Department of Forestry
Place Vincent Massey
351 St. Joseph Blvd.
Hull, PQ K1A 1G5
(819) 997-1107

REGIONAL OFFICES

Environment Canada
Atlantic Region
15th Floor, 45 Alderney Dr.
Dartmouth, NS B2Y 2N6
(902) 426-7990

Environment Canada
Quebec Region
CP 6060, 3 Buade St., 4th Floor
Quebec, PQ G1R 4V7
(418) 648-7204

Environment Canada
Ontario Region
25 St. Clair Ave. E., 6th Floor
Toronto, ON M4T 1M2
(416) 973-6467

Environment Canada
Western and Northern Region
2nd Floor, 4999 – 98 Ave.
Edmonton, AB T6B 2X3
(403) 468-8075

Environment Canada
Pacific and Yukon Region
Communications Directorate
3rd Floor, Kapilano 100
Park Royal South
West Vancouver, BC V7T 1A2
(614) 666-5900

****

Provincial and Territorial Governments

NEWFOUNDLAND

Department of Environment and Lands
P.O. Box 8700
St. John's NF A1B 4J6
(709) 576-3394

Department of Forestry and Agriculture
P.O. Box 8700
St. John's NF A1B 4J6
(709) 576-3245

NOVA SCOTIA

Department of the Environment
P.O. Box 2107
Halifax, NS B3J 3B7
(902) 424-5300

Department of Lands and Forests
P.O. Box 698
Halifax, NS B3J 2T9

NEW BRUNSWICK

Department of the Environment
P.O. Box 6000
Fredericton, NB E3B 5H1
(506) 453-3700

Department of Natural Resources and Energy
P.O. Box 6000
Fredericton, NB E3B 5H1
(506) 453-2614

PRINCE EDWARD ISLAND

Department of the Environment
P.O. Box 2000
Charlottetown, PE C1A 7N8
(902) 368-5280

Department of Energy and Forestry
P.O. Box 2000
Charlottetown, PE C1A 7N8
(902) 368-5010

QUEBEC

Ministère de l'Environnement
3900, rue Marly
Ste-Foy, PQ G1X 4E4
(418) 643-6071

Ministère de l'énergie et des ressources
200, chem. Ste-Foy
Quebec, PQ G1R 4X7
(418) 643-8060

ONTARIO

Ministry of the Environment
135 St. Clair Ave. W.
Toronto, ON M4V 1P5
(416) 323-4321

Ministry of Natural Resources
Whitney Block, 99 Wellesley St. W.
Toronto, ON M7A 1W3
(416) 965-2000

MANITOBA

Department of Environment
330 St. Mary Ave.
Winnipeg, MB R3C 3Z5
(204) 945-4742

Department of Natural Resources
1495 St. James St.
Winnipeg, MB R3H 0W9
(204) 945-6658

SASKATCHEWAN

Department of the Environment and Public Safety
3085 Albert St.
Regina, SK S4S 0B1
(306) 787-6113

Saskatchewan Parks and Renewable Resources
3211 Albert St.
Regina, SK S4S 5W6
(306) 787-2700

ALBERTA

Department of the Environment
Oxbridge Place
9820 – 106 St.
Edmonton, AB T5K 2J6
(403) 427-2739

Department of Forestry, Lands and Wildlife
9915 – 108 St.
Edmonton, AB T5K 2C9

BRITISH COLUMBIA

Ministry of Environment
Parliamant Buildings
Victoria, BC V8V 1X5
(604) 387-1161

Ministry of Parks
4000 Seymour Place, 3rd Floor
Victoria, BC V8V 1X5
(604) 356-7043

Ministry of Forests
1450 Government St.
Victoria, BC V8W 3E7
(604) 387-5255

NORTHWEST TERRITORIES

Department of Renewable Resources
P.O. Box 1320
Yellowknife, NT X1A 2L9
(403) 873-7420

YUKON TERRITORY

Department of Renewable Resources
P.O. Box 2703
Whitehorse, YT Y1A 2C6
(403) 667-5634

Earning The EcoLogo

The federal government, in conjunction with the Canadian Standards Association, has introduced its Environmental Choice Program to help consumers identify environment-friendly products that meet certain criteria. Products that have been certified by the Environmental Choice Board are then eligible to carry the EcoLogo emblem. To date, criteria have been established for the following product categories:

- re-refined lubricating oils
- miscellaneous products from recycled paper (hobby and craft forms)
- construction material from recycled wood-based cellulose fibre (thermal insulation)
- newsprint from recycled paper
- fine paper from recycled paper
- products from recycled plastic
- heat recovery ventilators
- batteries (zinc air)
- batteries (non-rechargeable)

- water-based paints
- solvent-based paints
- compost
- composting systems for residential waste
- energy-efficient lamps
- water-conserving products
- diapers (cloth)
- diaper services
- reusable shopping bags
- automotive fuels (ethanol-blended gasoline)

For more information, contact the Environmental Choice Program (107 Sparks St., 2nd Floor, Ottawa, ON K1A 0H3; (613) 952-9463). For a list of companies that have received the EcoLogo, or for copies of draft guidelines and to participate in the public review process, contact the Canadian Standards Association (178 Rexdale Blvd., Rexdale, ON M9W 1R3; (416) 747-4000).

As of March, 1991, here are the companies that had been licensed to use the EcoLogo:

* *

Automotive Fuels

Mohawk Oil
325 – 6400 Roberts St.
Burnaby, BC V5G 4G2

* * * * * * * * * * * * * * * * * *

Batteries
(Specialty/Zinc-Air)

Rayovac
5448 Timberlea Blvd.
Mississauga, ON L4W 2T7

* * * * * * * * * * * * * * * * * *

Composting
(Residential Waste)

Westman Plastics
19 Industrial Rd.,
P.O. Box 882
Dauphin, MB R7N 3J5

* * * * * * * * * * * * * * * * * *

Construction Material
(From Recycled Wood-Based Cellulose Fibre)

Can-cell Industries
16355 – 130 Ave. N.W.
Edmonton, AB T5V 1K6

Prosumex Insulation
820 Ellingham
Pointe Claire, PQ H9R 3S4

* * * * * * * * * * * * * * * * * *

Diapers

Altrim
450 Beaumont
Montreal, PQ H4N 1T7

Babykins
4 3531 Jacombs Rd.
Richmond, BC V6V 1Z8

Hapi-Napi Diapers
110 – 12220 Vickers Way
Richmond, BC V6V 1H9

Indisposable Cotton Diaper
491 Pacific Blvd.
Vancouver, BC V6D 5G6

Kooshies Diapers
336 Green Rd.
Stoney Creek, ON L8E 2B2

MED-I-PANT
4100 Parthenais St.
Montreal, PQ H2K 3T9

* * * * * * * * * * * * * * * * *

Diapers

Morgan Rapps
Bayside, R.R. 1
St. Andrews, NB E0G 2X0

Puritys Diaper
222 Newkirk Rd.
Richmond Hill, ON L4C 3G7

* * * * * * * * * * * * * * * * *

Fine Paper from Recycled Papers

Barber-Ellis
90 – 10551 Shellbridge Way
Richmond, BC V6X 2W9

Bowne of Toronto
60 Gervais Dr.
Don Mills, ON M3C 1Z3

Cascades
P.O. Box 2000
East Angus, PQ J0B 1R0

CPM
131 Sullivan St.
P.O. Box 1280
Claremont, NH 03743-1280

Dennison Manufacturing
200 Base Line Rd. E.
Bowmanville, ON L1C 1A2

Domtar
395 de Maisonneuve Blvd. W.
Montreal, Pq H3A 1L6

DRG Globe Envelopes
14730 12th Ave.
Edmonton, AB T5L 3B3

Gagne Printing
80 Ave. Saint-Martin
Louiseville, PQ J5V 1B4

Hilroy
250 Bowie Ave.
Toronto, ON M6E 2R9

IBM Canada
844 Don Mills Rd.
North York, ON M3C 1V7

Innova Envelope
56 Steelcase Rd. W.
Markham, ON L3R 1B2

Island Paper Mills
1010 Derwent Way
Annacis Island
New Westminster, BC V3K 5A5

Moore Business Forms
1600 – 130 Adelaide St. W.
Toronto, ON M5H 3R7

National Paper Goods
144 – 158 Queen St. N.
P.O. Box 2339
Hamilton, ON L8N 4E1

Noranda Forest Recycled Papers
67 Front St. N.
P.O. Box 1046
Thorold, ON L2V 3Z7

Prestonia Office Products
162 King St.
Stratford, ON N5A 6T1

Rolland
500 – 100 Alexis-Nihon Blvd.
Saint Laurent, PQ H4M 2P1

Southam Paragon
140 Boulevard de L'Industrie
Candiac, PQ J5R 1J2

Tenex Data
45 Commander Blvd.
Agincourt, ON M1S 3Y3

* * * * * * * * * * * * * * * * *

Lubricating Oil
(Re-Refined)

BresLube
625 Hood Rd.
Markham, ON L3R 4E1

Canadian Tire
2180 Yonge St.
P.O. Box 770, Station K
Toronto, ON M4P 2V8

Chevron
1500 – 1050 West Pender St.
Vancouver, BC V6E 3T4

Esso Petroleum
5 St. Clair Ave. W.
Toronto, ON M4V 2Y7

Hub Oil
5805 17th Ave. S.E.
Calgary, AB T2A 0W4

Loblaw
22 St. Clair E.
Toronto, ON M4T 2S5

Mohawk Oil
6400 Roberts St.
Burnaby, BC V5G 4G2

Turbo Resources
9830 34th St.
Edmonton, AB T6B 2Y5

* * * * * * * * * * * * * * * * *

Paints
(Water-Based)

Beaver Lumber
7303 Warden Ave.
MArkham, ON L3R 5Y6

Benjamin Moore
139 Mulock Ave.
Toronto, ON M6N 1G9

Betonell
8600 de l'Epee
Montreal, PQ H3N 2G6

Canadian Tire
2180 Yonge St.
P.O. Box 770, Station K
Toronto, On M4P 2V8

Chateau Paints
440 Ave. Beaumont W.
Montreal, PQ H3N 1T7

Color Your World
10 Carson St.
Toronto, ON M8W 3R5

Crown Diamond
3435 Pitfield Blvd.
St. Laurent, PQ H4S 1H7

ICI Paints
8200 Keele St.
Concord, ON L4K 2A 5

International Paints
19500 Trans Canada Hwy.
Baie D'Urfe, PQ H9X 3S8

Para Paints
11 Kenviw Blvd.
Brampton, ON L6T 5G5

Peinture UCP Paint
1785 Boulevard Fortin
Lavel, PQ H7S 1P1

P.S. Atlantic
69 Glencore Dr.
P.O. Box 546
Mount Pearl, NF A1N 2W4

Selectone Paints
39 Gail Grove
Weston, ON M9W 1M5

Sico
2505 De La Metropole St.
Longueuil, PQ J4G 1E5

St. Clair Paint and Wallpaper
2600 Steeles Ave. W.
Concord, ON L4K 3C8

Plastic Products Using Recycled Plastics

Du Pont
201 South Blair St.
Whitby, ON L1N 5S6

Superwood Ontario
2430 Lucknow Dr.
Mississauga, ON L5S 1V3

Further Reading

Chapter 1: Costing the Earth

- Brown, Lester R., et al. *State of the World 1990: A Worldwatch Institute Report on Progress Toward a Sustainable Society*. New York and London: Norton, 1990.
- Carson, Rachel. *Silent Spring*. Boston: Houghton Mifflin, 1962.
- Day, David. *The Eco Wars: True Tales of Environmental Madness*. Toronto: Key Porter Books, 1989.
- Samuel S. Epstein. *The Politics of Cancer*. San Francisco: Sierra Club Books, 1978.
- Lovelock, J.E. *Gaia: A New Look at Life on Earth*. Oxford and New York: Oxford University Press, 1979.
- Lovins, Amory B. *Soft Energy Paths: Toward a Durable Peace*. New York: Harper & Row, 1979.
- Moore-Lappé, Frances. *Diet for a Small Planet*. Revised edition, New York: Ballantine, 1982.
- Null, Gary. *Clearer, Cleaner, Safer, Greener: A Blueprint for Detoxifying Your Environment*. New York: Villard Books, 1990.
- Samuels, Mike and H.Z. Bennett. *Well Body, Well Earth: The Sierra Club Environmental Health Sourcebook*. San Francisco: Sierra Club Books, 1983.
- Schumacher, E.F. *Small Is Beautiful*. New York: Harper & Row, 1973.
- Seymour, John, and Herbert Girardet. *Blueprint for a Green Planet*. New York: Prentice-Hall, 1987.
- World Commission on Environment and Development (Brundtland Commission). *Our Common Future*. Oxford and New York: Oxford University Press, 1987.

✳✳

Chapter 2: Food and Drink

- Council on Economic Priorities. *Shopping for a Better World*. New York, 1989.
- Kneen, Brewster. *From Land to Mouth: Understanding the Food System*. Toronto: NC Press, 1989.
- Moore Lappé, Frances, and Joseph Collins. *World Hunger: Twelve Myths*. New York: Grove, 1986.
- Ontario Public Interest Research Group. *The Supermarket Tour*. Toronto, 1990.
- Robbins, John. *Diet for a New America*. Walpole, NH: Stillpoint Publishing, 1987.

✳✳

Chapter 3: Cleaners

- Dadd, Debra Lynn. *The Nontoxic Home*. New York: St. Martin's Press, 1986.
- *Favorite Helpful Household Hints*. [The editors of Consumer Guide.] Lincolnwood, IL: Publications International, 1986.
- Heloise. *Hints for a Healthy Planet*. New York: Perigee Books, 1990.
- Hunter, Linda Mason. *The Healthy Home: An Attic-to-Basement Guide to Toxin-free Living*. New York: Pocket Books, 1990.
- Wallace, Dan. *The Natural Formula Book for Home and Yard*. Emmaus, PA: Rodale Press, 1982.

✳✳

Chapter 4: Clothing and Toiletries

- Genders, Roy. *Cosmetics from the Earth: A Guide to Natural Beauty*. New York: Alfred Van Der Merck Editions, 1986.

****

Chapter 5: The Home

- Altman, Roberta. *The Complete Book of Home Environmental Hazards*. New York: Facts on File, 1990.
- Bower, John. *The Healthy House: How to Buy One, How to Cure a "Sick" One, How to Build One*. Don Mills, ON: General, 1989.
- Dadd, Debra Lynn. *Nontoxic and Natural: How to Avoid Dangerous Everyday Products and Buy or Make New Ones*. New York: St. MArtin's Press, 1984.
- Greenfield, Ellen J. *House Dangerous: Indoor Pollution in Your Home and Office — And What You Can Do About It!* New York: Vintage, 1987.

****

Chapter 6: Gardening

- Bennett, Jennifer. *Northern Gardener*. Camden East, ON: Camden House Publishing, 1982.
- Crockett, James Underwood. *Crockett's Victory Garden*. Boston: Little, Brown, 1977.
- Cullen, Mark. *A Greener Thumb*. Markham, ON: Penguin, 1990.
- Damrosch, Barbara. *The Garden Primer*. New York: Workman, 1988.
- *Harrowsmith Magazine*. Camden East, ON: Camden House.
- Editors of Mother Earth News. *The Healthy Garden Handbook*. New York: Simon and Schuster, 1989.
- *Organic Gardening Magazine*. Emmaus, PA: Rodale Press.
- Thompson, Bob. *The New Victory Garden*. Boston: Little, Brown, 1987.

ACKNOWLEDGEMENTS

Pollution Probe salutes all the people who helped put together the landmark first edition of The Canadian Green Consumer Guide; they laid the impressive groundwork for what has become the best-selling environmental guide in Canadian publishing history. We wish to acknowledge, in particular, the contribution of Gord Perks and Barbara Czarnecki in co-ordinating the scores of researchers and writers.

This time around, William M. Glenn and Randee L. Holmes were responsible for the research, writing, and co-ordinating. They were greatly assisted by Nancy McFarlane, who put in many long hours alongside them. There were also a host of volunteers and Pollution Probe staff who contributed comments, tracked down leads, and dug out obscure bits of environmental lore. In particular, we would like to thank:

Jennifer Beatty, Dave Bruer, Ross Campbell, Patricia Chilton, Robin Drever, Leslie Fleck, Mark Flemming, Larry Gaitskell, Emma Garrard, Marcus Ginder, Jennifer Good, Donald Holmes, Andrea Imada, Lori Mackay, David McRobert, Lori Miller, Joey Moore, Paul Muldoon, Tusher Pattni, Trish Ray-Taylor, Janet Sumner, Victoria Taylor, Charmaine Thompson, George Tierney, and the Ministry of Natural Resources (for funding through the Environmental Youth Corps program).

Probe also wants to thank the ad hoc review committee that pored over the first edition, updating it, expanding its focus, and looking for ways to make it even more useful to Canadian consumers. Our task would have been impossible without them.

Leslie Ballentine, Ontario Farm Animal Council, Jean Szkotnicki, Wholesome Food Council of Canada, and Jeff Wilson, Agricultural Groups Concerned About Resources and the Environment; Jake Books and Eugene Ellmen, The Social Investment Organization; Patricia Guyda, Canadians for Health Research; Linda Hawke, Toronto Humane Society; Jamie and Linda Holmes; Jill McWhinnie, Adam Blezensky and Irene Fedun, Recycling Council of Ontario; David Morris, agronomist; Thomas Nimmo, Canadian Organic Growers; J.A. O'Connor, The Canadian Chemical Producers' Association; Linda Pim; and Greg Simpson.

Finally, Pollution Probe is grateful to the hundreds of readers who have telephoned, written or dropped in to our offices over the past eighteen months with their suggestions and questions, congratulations and complaints. Their incisive comments formed the basis for our revisions.

Patrick Abram, Dynesco Inc.; Doris Azzopardi; Lanny Baer, Sanitation Equipment; Garry Belanger and Michael Smith, Duracell Canada Inc.; T.J. Carrothers, Javex Manufacturing Inc.; Christine Code, CMC Pedalogical Cycling Instruction; Ann Coffey, Lever Brothers Limited; Robert J. Dell, Dell Tech Laboratories Ltd.; Tom Donovan, Emerson Electric Canada Limited; Douglas S. Edwards, Canadian Carpet Institute; Tony Fairbrother, Bread & Roses; Angelina Ferreira; D. Mark Fresman, WCI Canada Inc.; Paul

Gibbard, Ontario Federation of Food Co-operatives and Clubs; Dr. Mikko Harri; Bob Hartogsveld, Ontario Ministry of Energy; Bob Hemp, Environment and Plastics Institute of Canada; Kathleen Hillborn; Jim Johnson, Canadian Renewable Fuels Association; Marina Kovrig, Recochem Inc.; Trish Krause, Burson-Marsteller; Peter D. Lawson, Canadian Water and Wastewater Association; Richard Lipman, Canadian Window and Door Manufacturers Association; Cheryl May, Absolutely Diapers! Inc.; Rob McMonagle, Solar Energy Society of Canada Inc.; Sharon Metz, City of Regina; K.J. Ott, Canadian Manufacturers of Chemical Specialties Association; Sara Perks; Dave Puskas, Dry Cleaners and Launderers Institute; Mark Sandercock; Don Spring, Photo Marketing Association International; R.Wm.C. Stevens, Poultry Industry Centre; W.E. Uren, Eveready; Basil Weedon, Canadian Aerosol Information Bureau; Peter Weinrich, Canadian Crafts Council; Ian R. West, S.C. Johnson Limited; Brian Wheeler, Canadian Battery Manufacturers' Association; Vivian White; Michael O'Sullivan, World Society for the Protection of Animals; Ainslie Willock, Toronto Humane Society.

Special mention must go to the revised guide's editor, Shaun Oakey, whose dedication, sound judgement, and patience were invaluable. At McClelland & Stewart, thanks are due James Adams, Martin Gould, Lynn Shannon, Trish Lyon, and Peter Buck.

INDEX

A MESSAGE FOR OUR READERS

Pollution Probe has been fighting to make this country a little greener for more than 20 years. Acid rain, hazardous waste, energy conservation, the quality of drinking water – each year we provide information on these and other vital issues to thousands of Canadians, from school kids working on projects to professional researchers writing reports.

It's a big, demanding job for a non-profit organization. If you and your family and friends wish to get involved in the work of Pollution Probe, please consider making a tax-deductible donation.

Name _____   City _____

Address _____   Province _____

_____   Postal Code _____

My contribution is: $_____   I am contributing by: Cheque ❏   Credit Card ❏

Card Name & No. _____   Expiry date: _____

Pollution Probe is a registered national charity (#0384750-53-13)
Donations, comments, and suggestions will be gratefully received at Pollution Probe,
12 Madison Avenue, Toronto, Ontario M5R 2S1 (phone: 416-926-1907; FAX: 416-926-1601)

Design, Photography, and Illustration: The Watt Group

With the following exceptions:

Illustration: Stephen Quinlan Illustration Ltd.
Pages 10, 22, 33, 40, 54, 57, 64, 70, 91, 94, 106, 108, 120, 127, 130, 137, 142, 145

Robert Meecham
Front Cover

Photography: The Ontario Ministry of the Environment
Pages 11, 12, 19, 28, 107

The Pollution Probe Foundation
Pages 14, 18, 21, 113

The Toronto Humane Society
Pages 20, 61

KAI Slide Bank
Page 24

Greenpeace
Page 34

The Image Bank
J. Ramey Page 140